In Search of **Tom Slick**

Courtesy of the Mind Science Foundation

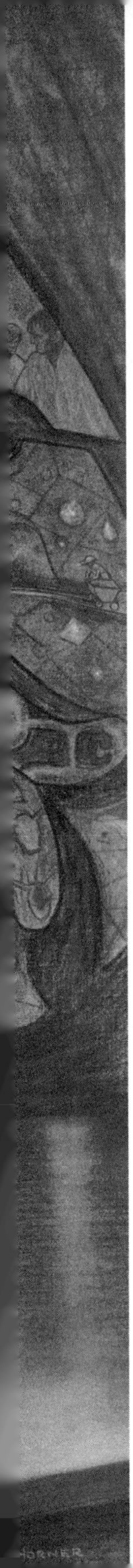

In Search of Tom Slick

Explorer and Visionary

REVISED EDITION

Catherine Nixon Cooke

Texas A&M University Press
College Station

Revised edition

This paper meets the requirements of ANSI/NISO Z39.48–1992 (Permanence of Paper).
Binding materials have been chosen for durability.
Manufactured in the United States of America

Library of Congress Cataloging-in-Publication Data

Names: Cooke, Catherine Nixon, 1950– author.
Title: In search of Tom Slick : explorer and visionary / / Catherine Nixon Cooke.
Other titles: Tom Slick, mystery hunter!
Description: Revised [and expanded] edition. | College Station : Texas A&M University Press, [2020] | First published by Paraview, Inc. in 2005 under title Tom Slick, mystery hunter! | Includes bibliographical references and index.
Identifiers: LCCN 2019044535 | ISBN 9781623498719 (cloth) | ISBN 9781623498726 (ebook)
Subjects: LCSH: Slick, Tom, 1916-1962. | Texas Biomedical Research Institute—Biography. | Southwest Research Institute—Biography. | Mind Science Foundation (San Antonio, Tex.)—Biography. | Businessmen—Texas—Biography. | Petroleum industry and trade—United States—Biography. | Explorers—Texas—Biography. | Inventors—Texas—Biography. | Philantropists—Texas—Biography. | Adventure and adventurers—Texas—Biography.
Classification: LCC HD9570.S59 C66 2020 | DDC 338.7/6223382092 [B]—dc22
LC record available at https://lccn.loc.gov/2019044535

For Betty Slick Moorman,
who asked her big brother to find the ghosts in the attic
(1917–2000)

Courtesy of the Mind Science Foundation

Contents

Courtesy of the Texas Biomedical Research Institute

Preface

Where No One Else Has Ever Been

When you embark for strange places, don't leave any of yourself safely on shore.
Be brave enough to live life creatively.
The creative place is where no one else has ever been.
You have to leave the city of your comfort and go into the wilderness of your intuition.
You can't get there by bus, only by hard work and risk . . .
. . . and by not quite knowing what you are doing.
What you'll discover will be wonderful; what you'll discover will be yourself.

—Alan Alda

When I heard actor and activist Alan Alda deliver the commencement address at his daughter's college graduation, I immediately thought about another graduate, a Phi Beta Kappa at Yale University more than seventy-five years ago, who lived his life in that wilderness of intuition, never leaving any part of himself "safely on shore." He was a risk taker who traveled to strange places in search of mystery; and Alan Alda was right—what he discovered was wonderful, in the truest sense of the word.

I knew Thomas Baker Slick Jr. as an uncle; others knew him as a millionaire wildcatter and explorer who drilled for oil and found it, searched for the Yeti in Nepal, spied for the OSS in World War II, discovered diamonds in the Amazon Basin, developed the Brangus breed of cattle and the Liftslab method of construction, established five major research institutes (the first when he was just twenty-five years old), wrote two books on world peace, and died a mysterious death at age forty-six. If it sounds like a movie plot, it almost was—*Tom Slick Mystery Hunter* was optioned by Elysium Films/Storyscape Entertainment in 2015, screenwriter Paul Boardman created an exciting script, and several well-known actors expressed interest in portraying its dashing and complex hero. As often happens in Hollywood, in 2018 the project stalled in development, but recently, several other production companies have indicated interest in bringing this story to television or the wide screen. Tom Slick's incredible persona and his search into the unknown provide a timeless tale . . . who knows what is ahead for it?

Uncle Tom was a favorite with all children, a Pied Piper cross between a magician and Santa Claus, who understood the importance of playfulness in life. When his spectacular home designed by O'Neil Ford was completed in 1960, I was nine years old. I still remember the wonderful housewarming party where guests ranged in age from five years old to ninety. One man wore a bright red turban; a glamorous woman wore chiffon and lots of perfume, and she was a big topic of conversation at breakfast tables the next day.

A fabulous sunken Japanese tub with fish swimming in it was in my eleven-year-old cousin Patty's bathroom; Uncle Tom told us that someday we would travel to exotic places and learn that there were all sorts of ways to bathe, eat, dress, and discover life. Outside, a motorized miniature Model-T Ford circled the driveway, with my ten-year-old cousin Tom at the wheel, his younger brother, Chuck, beside him honking the horn. With a wink, Uncle Tom confided that cars had an uncanny power to attract girls. I've often proven him right in later life.

After a dinner that included some strange courses like snails, and little bowls for dipping fingers, the conversation turned to new research underway at the Southwest Foundation (now Texas Biomedical Research Institute) and some additions that Uncle Tom had made to his growing art collection. I had never heard of Pablo Picasso, but the grownups were duly impressed. As we left the table, he led us down a hallway of contemporary art; our parents helped interpret the paintings for us, pointing out sunsets, ships, and landscapes amid the colors and shapes. At the end of the hall, Uncle Tom told us that some of the canvases had been painted by the chimpanzees at the Southwest Foundation! He suggested that we should each challenge the meaning of "art." All the children were delighted; the adults were not as amused.

Like everyone, Tom Slick was shaped by his family situation, his share of disappointments and losses, successes and extraordinary experiences. An early commitment to risk taking, and belief in the intuitive process, influenced his life in remarkable ways. His contributions to the twentieth century include breakthroughs in science, agriculture, architecture, and technology, and a long list of unsolved mysteries for us to explore in the centuries ahead.

What gave him the courage to listen to his intuition, journey to strange places, and live life creatively, as Alan Alda urged those college graduates to do in 1996? Was it his childhood experiences, like the death of his father, when Tom was only fourteen years old, or the much-publicized kidnapping of his stepfather by gangster Machine Gun Kelly a few years later? Was it the responsibility created by great wealth to contribute something to the world? Was it an inability to focus on just one thing, requiring large servings of variety and change? Was it the universally human search for "the holy grail," the hero's journey we each follow in our own way? What was the landscape like, in the deep wilderness of his intuition? What complex thoughts and feelings were hidden in the corners of his mind and heart?

He once appeared on a popular television series of the 1950s called *What's My Line*, where the audience tried to guess the true identity and accomplishments of each guest on a panel. I was thrilled that my handsome uncle was on television, talking about his expeditions to far-off places in pursuit of mysteries. Decades later, his face stares out from the covers of annual reports for the scientific institutes he established; there is a Tom Slick Chair in World Peace at the University of Texas; and Hollywood has courted this real-life Indiana Jones who trekked in the Himalayas in search of the Yeti. Will the real Tom Slick please stand up?

My discovery of hundreds of letters that Tom Slick wrote between 1948 and 1962, the year he died, filed in stacks of old boxes in a shed at one of his research foundations, provided me with some important clues. As I read them, I climbed into the head of a phenomenal thinker; the process made me wonder if the email correspondence that dominates our communication today will destroy those special insights so unique to letters. Interviews with his ninety-eight-year-old aunt, his sister and brother, his ex-wives, his four children, scientists, secretaries, oilmen, former lovers, and explorers like Sir Edmund Hillary gave me glimpses of a brilliant and complicated man. The "real" Tom Slick remains somewhat of a mystery, but his letters, adventures, and the memories of those who knew him best provide us with some interesting clues and inspiration as we each explore our own mysteries. I think he would like that.

In Search of **Tom Slick**

Mount Makalu, the world's fifth tallest mountain, and the Arun Valley below it, have beckoned to explorers for centuries. *Susan J. Taylor Photo Collection*

1

Ghost Call

April 2001

I hear birds screeching followed by the rumble of distant thunder. A thick mist surrounds me, and it is very difficult to see. Ice is falling from the sky like sharp shards of glass; and I am very cold. I can barely discern a track of some sort, ghostly footsteps in the snow . . .
I follow.

The dream has come again, both scary and seductive. I zip my sleeping bag so that just my face is exposed, in a darkness so complete that the roof of my tent is invisible. I listen. Yes, the thunder is real.

A ghost has lured me to Nepal. For the past two years, his pale blue eyes have stared at me from a small silver-framed photo; his correspondence, now yellowed with time, has given me entry to a private world of dreams and visions. Images have haunted me—snowy mountains, hidden valleys, dark and sacred caves, and mysterious creatures. I could not resist the call to his mythical Shangri-La; and like everyone who searches for dreams, I believe I might find answers just over that farthest horizon.

A group of six friends, boldly calling ourselves "twenty-first-century explorers," has organized our quest—an expedition to Nepal's most remote region, where mountain gods still control the rain and snow, and hidden *beyuls* (valleys) hold the promise of grail-like discovery. East of Mount Everest, in the shadow of Mount Makalu, the world's fifth-highest peak, we have planned to trace the route once taken by the blue-eyed "ghost" nearly half a century ago. He was searching for the mysterious Yeti, or Abominable Snowman, whose legend has inspired expeditions to the tiny Himalayan kingdom for more than one hundred years. We will not only look for the Yeti, but also for the ghost of Thomas Baker Slick Jr., determined to discover how mysteries guide our life journeys, and why the unknown pulls all of us like a powerful magnet.

Modern explorers like Daniel Taylor and mountaineer Reinhold Messner have made similar searches, and their books describe glorious, though difficult, adventures. There still are occasional reports of Yeti sightings, glimpses of a creature crossing the high snowcapped ridges or roaming the isolated thick forests of the Arun Valley. Messner has decided it's a chemo, a member of the bear family; Taylor still thinks that a new species might exist in these mountain ranges, as did Peter Matthiessen, the award-winning author of *The Snow Leopard*. In 2019, the Indian Army filed an official report that officers had seen Yeti footprints in the remote Arun Valley, and photos published in media including the *Washington Post* sparked new excitement. So, despite the skepticism of most scientists, the mystery and possibility of the Yeti endure.

We have brought expensive night vision equipment—infrared binoculars and cameras—for our adventure in Nepal. State-of-the-art satellite maps have helped us chart our course over a terrain where there are virtually no existing trails and very few inhabitants. We have *not* brought along any communication equipment—no satellite cell phone, no laptop powered by a portable generator, not even a radio. For the next three weeks, we will be in a faraway world, unable to hear the news and noises of our Western lives, experiencing a frontier of mystery the way Tom Slick did so many years ago.

When Tom Slick first journeyed to Nepal in 1956, he was forty years old, prematurely silver-haired, with a legendary twinkle in his blue eyes. His body was lean and fit for an entrepreneurial oilman who traveled often and enjoyed fine food and entertaining. But aside from a rather luxurious African safari in 1952, where the guests stayed in large canvas tents with comfortable beds and beautiful rugs, and many men's hunting and fishing trips, Slick's outdoor camping experiences were few. He had gathered an impressive expedition team for his search for the Yeti, including a well-respected big game hunter and adventurer named Peter Byrne, and Nepalese porters who found the steep ridges and dense forests as familiar and comfortable as a Westerner's "walk in the park."

We have done the same. Our expedition members all are fit by Western standards, but no one has climbed much above 12,000 feet before, and the thought of no bath for three weeks is a novel idea. Can we endure the physical challenges; will two months of endurance training be enough for the constant ups and downs of peaks and *beyuls* in the little-explored Upper Arun Valley? Can we put aside our Western mindsets to look for the serenity and magic that might await us in the Himalayas? Will there be any evidence of the Yeti, the elusive creature that drew Tom Slick to the Himalayas three times in the late 1950s? And finally, what will Slick's "ghost" reveal to us about himself, and the many hidden mysteries that wait for us just beyond the ranges?

With a special expedition flag awarded to our group by The Explorers Club, the international organization established in 1904 to promote

The 2001 Arun Valley Expedition team carried the prestigious Explorers Club flag; *left–right*: Susan Taylor, Dr. Brad Towne, Rima Lewis, Dan Bennett, Catherine Nixon Cooke, and Brian Hanson, near the base of Mount Makalu. *Catherine Nixon Cooke Photo Collection*

exploration and scientific contribution all over the world, expert route planning by Scott Hamilton, whose medical expeditions to Mount Everest have yielded valuable data about the effects of high altitude on our bodies, and Bibhuti Chand Thakur, manager of Himalayan Holidays Trekking in Kathmandu, we feel "official." In a world where there are no coincidences, "Bibhu" tells us that his father was the Nepalese army officer who accompanied Slick on his expeditions. Our confidence is boosted further by Rabi Thapa and Ram Sharan Thapaliya, two guides who have visited the Arun Valley once before, more than twenty-five porters, the latest in hiking boots and expedition clothing, cameras, and of course, our excitement.

Three of the team members belong to The Explorers Club: Brian Hanson and I have served on the organization's board of directors, and twenty-eight-year-old Rima Lewis is a new member of the club, with a recent law degree and training in mountain rescue. Dan Bennett is CEO of Sunbelt Sportswear, a large international women's sportswear mail order business; he has researched trekking equipment and brought along the very best, including a water filter system that will prove very valuable. Dr. Brad Towne is a physical anthropologist, an associate professor at the Lifespan Health Research Center at Wright State University Medical School, and an associate scientist at Southwest Foundation for Biomedical Research, a research institute founded by Tom Slick in 1941. He has done previous research in Nepal and is familiar with primate behavior and Yeti legends. And finally, Susan Taylor, a television journalist, and my best friend, has joined our team, leaving behind a busy career and family life for this adventure. A terrible chest cold has almost prevented her participation; her physician finally gave in to her sheer determination, equipping her with a small pharmacy for the adventure.

The sound of the big Russian helicopter is amazing, first affecting our ears and eventually vibrating through every part of our bodies. We reach for our earplugs and grip the narrow metal benches as the chopper lifts off, leaving dusty Kathmandu behind. Our backs to the windows, we face a huge pile of duffel bags and equipment in the middle of the helicopter's floor, turning uncomfortably to look out the small Plexiglas windows. Soon we've climbed to 9,000 feet and are skirting the foothills of the Himalayas. Nepal has the most dramatic elevation

A Russian helicopter flew us northeast from Kathmandu to the tiny village of NagiTar where porters and supplies were waiting for us. *Susan J. Taylor Photo Collection*

changes in the world—from nearly sea level in the south to the towering peaks of Mount Everest, Kanchenjunga, Lhotse, Makalu, and Chamlang, ranging from 29,000 to just over 24,000 feet high, respectively. As we bounce over air pockets, heading northeast, incapable of talking over the noise, each team member thinks about the days ahead. I remember my dreams of the ice storm, but the bright blue sky and sunshine soon chase those thoughts away.

After an hour of flight, Rabi enters the cockpit to talk to the morose Russian pilot. It turns out he has never flown to NagiTar before and is not quite sure where to land. It is a tiny village, just above Tamku at 6,500 feet, far enough from the populated Everest region that the villagers have never seen a helicopter. We touch down in a different world.

Ram is waiting on the ground with about twenty-five porters and a nurse practitioner named Paru Gaire. They have walked four days from Tumlingtar, where there is a tiny airstrip, carrying all of the tents, food, and kitchen equipment, and our camp is already set up in the flat meadow just below NagiTar.

Before lunch we hike to the small monastery that is nestled in terraced fields of potatoes, wheat, and mustard, surrounded by neat wooden compounds where fewer than one hundred people compose the population of NagiTar. It is a steep climb, but not very far. However, I'm aware that this easy thirty-minute journey has challenged my breathing, and I imagine what might be ahead in the days to come. When I reach the top, I smile at Susan, who also is panting just a little bit. She tells me I have blood on my teeth, and I discover I am having a nosebleed . . . not an auspicious start.

Descending through the green terraced fields, I'm struck by the spectacular beauty that surrounds us. Past journeys to the turquoise waters of the Mediterranean, the lush jungles of Africa, the white and craggy cliffs of England, the burnt orange desert of the American Southwest, and the dense green rice paddies of Indonesia have touched some part deep within me, thrilled me with their visual and spiritual power. But these mountains in the distance, with their snowy peaks hidden by mist, and these pristine valleys where gentle-eyed Nepalis look at us with curious eyes, create a physical reaction in me that comes as waves of awe, and wonder, and yes, some fear.

Dawa Tshering Sherpa was a porter on the Slick Expedition; he told us about capturing a baby bear and warned us that Bankatis (wild women) lived in the forest. *Susan J. Taylor Photo Collection*

Later, after dark and after dinner, we sit around a kerosene lamp and wait for our first visitor from the past, a sixty-three-year-old porter who accompanied Tom Slick on his expeditions in 1957 and 1958. Dawa Tshering Sherpa has walked for about forty-five minutes without light of any kind, barefoot and wearing a light jacket. Ghostlike, he appears at the entrance of the dining tent, just slightly over five feet tall, weighing less than a hundred pounds, and still very strong. He accepts a cup of tea and lets his memories speak.

It was a long time ago. I am an old man now and most of my friends have died. But I am living longer because I am small in stature. I am very strong—it was my job on the Slick Expedition to carry the tents. They were much heavier than the tents you have now because they were made of canvas.

We were looking for the Yeti near a waterfall close to the village of Wallung, where you will go in a few days. Some of the expedition members also wanted to find a Bankati ("wild woman" or "forest girl") that everyone knew lived there, but they couldn't find her. Instead they captured a baby bear.

There are no Bankatis close to NagiTar, but there still are some on the other side of Sedua, a few ridges over that way, where you will also go. Bankatis are very wild; they have very long breasts, which they flip over their shoulders when they run. They turn over rocks and eat frogs. I have never seen a Bankati, but I have seen the frog remains that they leave on the rocks in the forest. And I have heard the sounds they make.

Heeeeeeee … … … … heeeeeee!

An eerie high-pitched wail echoes in the quiet darkness, and the young porters from our expedition gather around the tent, wide-eyed, to listen to the old man. He tells us about the strange hunting dogs that the Slick Expedition brought to Nepal, imitating their howls, and admits that he ran away from the camp rather than following the cook's orders to kill a goat, something that he could not do as a Buddhist. He does not read or tell time, but he wears a watch that his son has sent him from India, and around 9:00 p.m., without even stars to guide him, he starts his walk home in total darkness.

The next morning, the trek begins. The next days are spent walking up steep ridges, often at a 75–80 degree angle, for hours, then descending rocky slopes, without a trail, for hours. Soon, all semblance of civilization (for that is now how we remember tiny NagiTar) has disappeared and we are chopping our way through bamboo forests, bushwhacking, heading toward the distant Mount Makalu, which is now hidden by the mist and rain that will become our constant companions.

The next week is spent feeling wet, cold, and tired; our team usually struggles into camp at nightfall, having walked anywhere from six to eleven hours. The physical challenges are occasionally forgotten as we glimpse sacred chortens, draped with prayer flags, on the top of seemingly untraveled ridges, and giant rhododendron trees, more than thirty feet tall, bursting with scarlet blooms. Every day is filled with suffering and redemption.

Toward the end of our first week of trekking, we approach Deurali Pass on a narrow trail along the ridgeline, with occasional glimpses of giant magnolia trees, blooming with white flowers. We begin our steepest ascent so far, up and over the pass at 10,300 feet. It is misty and cold; small snowflakes and faded fragments of prayer flags have accumulated on an ancient *Mani* wall. We walk it, in the traditional clockwise manner, placing flowers we picked along the trail on its craggy top, respectfully asking for a safe and rewarding journey.

Crunching through shallow snow, we traverse two more ridges, descending to Rat Mate, which is just a small flat clearing with two *goths*, rough wooden sheds that are used to protect yaks and herders during the winter. Our tents are erected, almost touching, and the porters light fires in the *goths*, using one for our kitchen and the other for their shelter. Soon drumming and singing join the sound of thunder. As a deluge of rain pounds our tents, we hear a Nepali "concert" a few feet away. It is better than a five-star hotel.

At five o'clock in the morning, an hour before our usual "wake-up" call, I hear a thumping on my nylon tent. It is Ram, knocking to tell me that the rain and fog have disappeared, and we have a rare view of Mount Makalu and Mount Chamlang that won't last long. Scrambling into my parka, I crawl out and look toward the mountains. It is breathtaking. Slivers of gold, that may eventually become a sunrise, or may be reclaimed by mist, frame the peaks. Were they really there yesterday, just beyond our vision? It is easy to imagine the mountain gods watching, sending thunder and rain yesterday, and giving us a glimpse of paradise this morning.

Rain and very few flat places made it difficult to position the tents; our team knew that Tom Slick's expedition had experienced the same challenges. *Brian P. Hanson Photo Collection*

After breakfast, we start toward the distant peaks, and the mist and rain have returned. There are more long days of exhausting ups and downs, made more difficult by the extreme weather. Now we wonder if the mountain gods are displeased with us. The rain continues, and at an altitude of 11,000 feet we encounter more snow. Up and down. Up and down. We collect samples of animal scat in special bags that Brad Towne has supplied, and we carefully set up our infrared cameras at night, although the rain makes it impossible to see anything. Where is the Yeti? Where is Tom Slick?

We camp at MaganTar, a small clearing with three *goths* and a beautiful old *chorten*, where members of the Slick Snowman Expedition set up a watch post in 1959. The next morning, sixty-three-year-old Ang Norbu Sherpa joins us for breakfast. Now a yak herder, he was a young porter for Slick and remembers that this site was headquarters for the Yeti hunters for three months. His job was to feed four baby bears that the Byrne brothers had captured, using a crude bottle with a mixture of yak milk and berries. The old porter tells us that the bears were eventually taken back to Kathmandu and shipped to a zoo in the United States, and that another member of the expedition team was a Bengali taxidermist who collected species of birds to send back to India.

When asked if he believes in the Yeti, the yak herder claims he has seen footprints at higher altitudes, near *goths*, and he tells us a story that we will hear several times during our trek. It has become part of Nepal's folklore.

A monk was meditating in his hut when he heard a banging on the door. But because he was in meditation, he could not open the door. The Yeti put its hand through the window and the monk saw that the creature had a thorn in its finger. The monk removed the thorn. The next day, there was a freshly killed deer at the monk's doorstep.

"Sometimes the Yeti repays a kindness with an act of kindness in return," Ang Norbu Sherpa tells us. "But if you hear a Yeti sound, you should not answer, because it might think you are one too. We don't fear the Yeti, but we are careful."

After photographs and a small gift of one hundred rupees and a silk kata, we leave the porter-turned-yak herder and begin our trek toward Dobatok and the sacred Khempalung Cave. On the twelfth day of our expedition we reach Dobatok, the halfway point of our journey. It seems a virtual metropolis, with six tiny homes and a beautiful prayer wheel. Again, we squeeze our tents onto a small bit of empty flat land, and we prepare for a very special experience the next day.

The Khempalung Cave is one of the legendary holy places that are said to have been hidden by Guru Rinpoche in the Himalayas centuries ago. It is universally worshipped—by Sherpa, Rai, Khumbo, Shingsapa tribes, and Tibetans of Kharte and Dingri. In her chapter in *Mandala and Landscape*, researcher Hildegard Diemberger describes the slopes of the *beyul* ("hidden valley") where the cave is located as "covered with thick forest, inhabited by wild animals, and surmounted by the white Himalayan peaks . . . in the *beyul* lives the snow lion, the plants know no seasons, there are fruits and flowers and many streams, there is a poisonous burning lake, there are medicinal springs, clouds and mists darken the whole atmosphere, there is a lot of rain; and from the sky you hear the roar of the dragon (thunder)."

During the past days, we certainly have heard the dragon and felt the rain, seen the clouds and mist, been amazed by the plants and flowers blooming almost magically from towering trees, and glimpsed the white peaks during brief moments of clear weather. Tired and more than a little giddy, I am sure the poisonous burning lake will be just around the next bend.

But our climb to the cave proves uneventful, and we are greeted by more than one hundred once-colorful prayer flags waving in a blur of faded red, green, blue, yellow, and white. They surround a simple stone structure with aluminum siding and wooden benches, where we remove our hiking boots and anything else made of leather. Rabi and Ram invite us to say a prayer and light butter lamps before moving outside to wash our feet in a sacred waterfall. We move down a rocky slope—painfully, since we are barefoot—to the entrance of the cave that Tom Slick described as an oracle. Perhaps his ghost will be waiting inside.

Just inside the rock ledge, there are paintings of the ancient Tara, the feminine goddess said to watch over the cave, as well as a rusted trident and bell. The trident represents the masculine; the bell is symbolic of the feminine; and both are revered in this holy place. Despite the

Ten-year-old Pema Sherpa appeared like a magical fairy to lead us through the sacred Khempalung Cave. *Brian P. Hanson Photo Collection*

warnings we'd read in academic journals that the cave is not for anyone claustrophobic or plump, I'm astonished at the smallness of the opening and I wonder if our larger Western bodies will have difficulty passing through the narrow chambers. A beautiful little girl named Pema will guide us through; she tells me she is ten years old and that the cave is her oracle. I think about the very different activities that my daughter, now in college, pursued on the other side of the world when she was Pema's age. I take a deep breath, and with our headlamps turned on, we crawl into a dark, very tight space.

Most of the journey is spent either moving along on all fours or slithering like a snake—using just the arms to pull the body along. Occasionally, we reach a chamber big enough to achieve a sitting position, and there is always a stone Buddha figure or other carved deity to greet us there. One can leave a coin or butter lamp or some other offering, along with a prayer for health, wealth, fertility, abundant crops, or whatever blessing one hopes for. I am simply hoping to see daylight again.

We reach a particularly tight passage that is called the Purification Chamber, where pilgrims must slide through a narrow, tube-like passageway, circling a rock column clockwise. At one point, when my body is in the shape of the letter "C," panic seizes me. I remind myself that I have come halfway around the world for this experience; I say my mantra softly to myself; I am calmer. I begin to slither along the rocky path; suddenly, the idea that I might be stuck overtakes me. I nearly scream, recognizing that I cannot stand up—I can't even sit up or raise my head! I feel myself sinking into a darkness far more frightening than the cave's blackness . . . then the tiny hand of Pema, who has gone through the Purification Chamber just ahead of me, reaches back to touch mine. I slither on.

I pull out of the C-shaped tunnel and hear the other team members decide not to pass that way. Later, after we've left the cave, Susan tells me that a huge spider had been overhead, just inches from me, during my entire "purification" effort.

We move farther into the cave, crossing a cold, shallow stream that is somehow comforting because it takes my mind off the narrow walls that confine me. The entire traverse takes forty-five minutes—forever—and I crawl into daylight with a renewed sense of life's gifts. According to Buddhist texts, three trips through the Khempalung Cave will bring a pilgrim sainthood. One-third saint is enough for me.

Just an hour's walk from the cave, we cross terraced fields to arrive in Gongtala, a village of about twenty houses constructed of mud and rough wood, with bamboo mat roofs. When we visit the home of a family whose father had been one of Slick's porters, we discover a treasure—an old cargo box, stenciled with Slick Snowman Expedition. We purchase the box for the asking price—500 rupees, about seven dollars —to take back as tangible evidence of the past we have come to explore. In celebration of their sale, the family presents us with a huge jug of *raksi*, a strong alcoholic drink made from barley or millet. I take a small sip to be polite, marveling at the delicate china cups that materialized for this special occasion, then feel the powerful drink burning my throat and exploding behind my eyes.

After a three-hour hike back to Dobatok, we discover heaven. Our porters have erected a "shower tent," and each team member is allocated a large basin of hot water and a pitcher for pouring the water over ourselves. This is our first—and only—bath of the journey. Our guides warn us that tomorrow will be a harder trek. The village of Wallung will be our destination, and we must be careful there—inhabitants have been known to poison visitors to acquire their power.

Leaving the hidden valley behind, our journey the next day is one of the hardest on the trek. We walk all morning in heavy rain, and the porters are having real trouble because there is no trail to follow. We are bushwhacking through thick forest and bad weather. By lunchtime, we realize that we have not made enough progress to reach Wallung before nightfall; everyone is cold, wet, and miserable. *Na ram-ra* ("not good") becomes a chant from the porters; we agree with them.

Up a ridge, down to a valley, there is much slipping and falling, and a horrible discovery that we are not alone. When I stop on a flat stretch of soggy ground, removing my backpack to find an energy bar, shedding my rain jacket for a moment, I cause a panic within our group. I look as if I've been shot by a machine gun, with blood splatters all over my t-shirt, places where leeches have attached under the warm band created by my backpack strap. We remove them by lighting matches, shivering, realizing there will be more as we struggle toward Wallung.

By the time we approach the village, it is dark; villagers come up the

Approaching the village of Wallung, at nearly 9,000 feet, the group was surprised by blood-sucking leeches in the dense forest. *Brian P. Hanson Photo Collection*

trail with flashlights to help us over the rocky slope to a small schoolhouse. Instead of going to our wet and leech-infested tents, we decide to sleep on top of the tables in the schoolroom, warm and dry in our sleeping bags. After removing more leeches, spooked by their bloodlust, we are exhausted. We agree that this day has been too much for us; we're not sure we can endure another one like it.

Apparently, our porters feel the same way—one says the day has been harder than the trek from Makalu Base Camp to Everest Base Camp; quite a few threaten to quit and go home.

We awaken to sounds of a loud argument between Rabi and the porters, who say the route is too difficult, especially with the terrible rain. Rabi redesigns the route; instead of going all the way to Sedua, we will take a shorter but steeper trek to Norbugaon, and he agrees to pay the porters more to continue. About half decline the offer and start their journey home.

Before leaving Wallung, we interview Dorji Sherpa, who participated in the 1958 Slick Expedition and is now a lama. He tells us he believes in the Yeti, "but it is very difficult to see. It is like a god; it can disappear at will."

Also living nearby is Tashima Sherpa, the seventy-two-year-old widow of another Slick Expedition team member. She tells us she is skilled in herbal medicine and magic potions; when she asks if we would like some refreshment, we remember Rabi's stories of poison in Wallung and politely decline. It is time to push on, following the modified route that will hopefully create an easier day of trekking.

After crossing two very steep ridges, at times clinging to the face of the mountain, we wonder what the more difficult day would have been like. Finally, we arrive in Norbugaon around five thirty, after just slightly more than eight hours of trekking. After de-leeching ourselves and unpacking our sleeping bags, we are almost too tired to eat dinner. Susan's chest cold has worsened, and Rima begins to feel sick at her stomach and excuses herself during the meal. Suddenly a strange woman appears in the doorway of the dining tent.

She screams and seems to be possessed by some unhappy spirit; dancing in a trance, whirling, shrieking. Rabi and Ram usher her away. Everyone goes to bed uneasy; this is not a peaceful place. Tomorrow we will travel to Tashigaon, which translates as "prosperous place," but to get there, the porters will have to bushwhack over slippery terrain. We are promised that after Tashigaon, the remainder of the trek will be well-established as we approach the Base Camp of Mount Makalu.

The next morning, we are picking off leeches again as we traverse a high ridge in the pouring rain, creating a trail where none exists, descending to Tashigaon by midafternoon. It is the first real village we've seen in weeks, the place where trekkers and climbers begin the Makalu route that takes them over Shipton Pass to Base Camp and on to the summit of the world's fifth-highest mountain. There is a small general store, and an "inn" with two bedrooms.

Some of us opt to sleep in the bedrooms, which each have four wooden cots where we place our sleeping bags; others prefer their tents outside. Just before we blow out the candle that lights the room, Paru removes a huge spider from just above my bed. Being Hindu, she cannot kill it, so she captures it with a Kleenex and gently takes it outside. I take a Tylenol PM . . . why are these spiders pursuing me?

The next morning, as I wash my face in the small aluminum bowl of hot water, I realize that I've come to consider this a luxury of the day. It startles me when Susan announces that there are only four more days left in our journey; just across snow-covered Shipton Pass, a helicopter will come for us, returning us to the modern lifestyles we left behind nearly a month ago.

The trek to Khongma is nearly all uphill, going from about 7,500 feet to 12,000 feet. For the most part, I feel strong and enjoy the workout, glad that we're going up, since going down has become increasingly painful for my knees. After about five hours of steady uphill, we encounter a terrible surprise. A freak ice storm, fast, furious, and very painful, attacks us. It is my dream, my nightmare, come true.

The slope is very slippery, of course, and to see where to go next, we must look up into shards of falling ice. I stop under a rock shelf to add another layer of clothes from my backpack, but I am already soaking wet and my fingers are too numb to work the zipper. No one is talking; we need every bit of breath and concentration to keep going. The mood is very serious as some of the porters turn back toward Tashigaon, others remove their rubber shoes because they are slipping too much in the snow and walk barefoot.

The expedition encountered snow in the high passes, and a freak ice storm made the climb to Khongma more challenging. *Brian P. Hanson Photo Collection*

Around three thirty in the afternoon, we nearly crawl into Khongma, which is just a flat shelf of snow, with one small aluminum hut with a kerosene stove, where a lady is making tea. More than half a century ago, Tom Slick camped on this same flat shelf and looked out at a brilliant blue sky and the imposing face of Mount Makalu in the distance.

We shiver in our wet clothes, drink tea, try to pretend that the kerosene stove emits heat, and pray for a blue sky tomorrow. Susan's cough sounds like pneumonia, the blizzard of snow turns to torrential rain, and the porters finally begin to struggle in with their sixty-pound packs, still barefoot.

Before pitching the tents, they must get warm too. Drums and *raksi* appear in the hut; the porters drink and laugh, despite the hardships of this day. Soon, the body heat of more than twenty people in a tiny space creates some warmth. At last, around eight-thirty, the final loads arrive in the darkness. We slip through rain and snow to wet tents, open our wet equipment bags, unpack our wet sleeping bags, and shivering, crawl inside. Everyone is too tired to eat.

The next morning is completely gray, with no visibility because of snow and fog. An Italian expedition team that also has camped at Khongma leaves by 6:00 a.m. for Shipton Pass, wearing crampons and carrying ice axes, equipment that our group has not brought with us. Our departure is delayed; Rabi recommends that we stay in Khongma another day, worried about Susan's congestion, Rima's dysentery, and the exhaustion of everyone in the group . . . not to mention our lack of proper equipment and gear.

Midmorning, grim news arrives from higher up the mountain that two of the Italian climbers have fallen while crossing Shipton Pass. The porter reports that the ridge is especially treacherous. We decide that the high passes ahead—Shipton-La, Tutu-La, and Keke-La—are too challenging without crampons for the unexpected ice and snow; wisdom tells us that we are not to be welcomed by the mountain gods this trip. One of our strongest young porters, Dawa Sherpa, is dispatched back down the mountain to Sedua, the nearest village with radio contact. It will be a day's run for him but would be a journey of several days for us. Dawa will radio our chartered helicopter, instructing it to pick us up in Tashigaon rather than in Yangle Kharka as originally planned.

We spend the rest of the day and the next night in our tents, emerging only for meals, with the wind blowing ferociously. I doze and listen to my portable Walkman, lulled by Fleetwood Mac singer Stevie Nicks's mysterious tales of witches and advice to "don't stop thinking about tomorrow." My sleeping bag is finally dry, and I sink in and out of dreams in its cozy warmth. All of us think about real mountain climbers who spend weeks like this, waiting for the weather to clear, hoping to summit their mountain.

The next morning, the mountain gods send us a message of approval. For the first time in weeks, the day is absolutely clear. There is not a cloud or snowflake or raindrop in the sky. The view of the mountains is incredible . . . it is hard to believe they were there yesterday, hidden by the snow and mist.

This is the view Tom Slick saw in 1956, 1958, and 1959. Not far from this spot, he found a footprint in the snow that he believed belonged to a Yeti. The plaster cast of the print that he made that day near Khongma would join his collection of Picassos and other treasures that adorned his home in Texas. The mysteries he had encountered were the same ones we were thinking about as we waited to hear the powerful helicopter blades in the mountain air.

The oracle at Khempalung Cave, the waterfall where the *bankati* still shrieks at night, the footprints we photographed—perhaps of an elusive snow leopard—the leeches and spiders . . . Tom Slick passed this way first, answering some mysterious call to explore, to search, hopeful that he would discover the creature rumored to be the "missing link" in human evolution. Now, so many decades later, we have undertaken his quest and we find that nothing has changed.

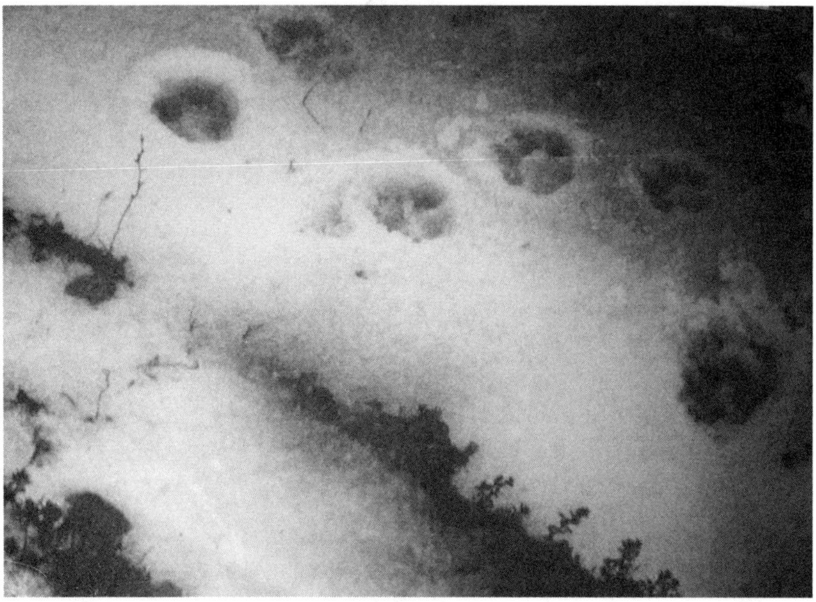

Not far from Shipton Pass, we photographed tracks of a rare snow leopard.
Susan J. Taylor Photo Collection

Margaret Urschel Photo Collection

2

Missing Links
1900–1930

In early May 1916, the King of the Wildcatters was waiting for two important events. One would occur in an oil field in northern Oklahoma, not far from his discovery of huge oil deposits in the Cushing Field in 1912; the other would occur in a hospital in Clarion, Pennsylvania, where his wife of just one year was about to give birth to their first child.

Thomas Baker Slick Sr. was a workaholic and an explorer; and his discovery of major oilfields in Oklahoma, Texas, and Kansas had earned him the nickname of "Lucky Tom" by 1912; and by 1929, he was the

Huge oil discoveries in Oklahoma in 1912 earned Tom Slick Sr. the nickname "King of the Wildcatters."
Betty Slick Moorman Photo Collection

largest independent oil operator in the world. As he waited for a new baby and a new oil well, both due the same week, he grappled with priorities, as he always had.

Complex emotions were familiar terrain to Slick Sr., and his thoughts raced from the almost soothing, practical checklist he followed when drilling a new site, to his wife, nearly one thousand miles away in Pennsylvania, about to give him a son and namesake. Former oil partner Bernard B. Jones had introduced Slick to Berenice Frates just over a year ago, and the couple married almost immediately. The thirty-two-year-old Slick wanted roots. Twenty-five-year-old Berenice Frates, oldest daughter of a railroad tycoon named Joseph Frates, was intelligent and attractive. She also understood the wanderlust spirit of her own father and presumably of the wealthy oilman who was courting her. She was a serious girl, with dark hair and eyes, who had grown up with an incredibly beautiful, vivacious mother, a gregarious, entrepreneurial father, and five younger siblings. As Joseph Frates worked his way up in the railroad business, the family traveled constantly, and Slick felt confident that his own gypsy spirit would be understood and tolerated by his new in-laws.

Before meeting Berenice in 1915, Slick had recognized that his work habits were not healthy. Following his legendary "Wheeler Strike"—the first oil well drilled in 1912 in the famous Cushing Field near Bristow, Oklahoma—he quickly purchased the majority of leases in the entire area, thanks to a clever scheme. Slick rented every transportation device within miles so no one else could reach the landowners. The field proved to be immense, eventually producing 310,000 barrels of oil a day at its peak.

Exhausted by the intensity of his work, and by the newfound success, he put his mother, sister Flored, and older brother Jesse in charge of his business and took an extended trip to Africa and the Orient. From rare letters to his family during the trip, it is obvious that Slick was tired and depressed; conditions that would haunt him all his life. Notorious for chain smoking and forgetting to eat when working on an oil deal, Slick realized he was ruining his health and wrote to his mother in 1912:

> *I am going to get well no matter how long it takes or how much money I lose. It always makes me sad when I think of these things while writing and it hurts me.*

After discovering the Cushing Field in 1912, Tom Slick Sr. took a trip to China. *Betty Slick Moorman Photo Collection*

He traveled alone for eight months to some of the most exotic places in the world. To document the adventure, he had his photograph taken in a Chinese rickshaw, one of the few times he posed for a camera. His worries that the trip might have a negative impact on his business were unfounded. In fact, while he was away, Prairie Oil and Gas bought some of his oil leases for one million dollars, an unbelievable sum in those days. Other leases had been developed by his brother, Jesse, and were producing large amounts of oil. When he returned to Oklahoma in 1913, he reentered the oil business with new gusto, and he accepted his friend and former partner Bernard B. Jones's matchmaking offer with equal enthusiasm.

A man who always trusted first impressions, Slick proposed to Berenice Frates soon after they met, and the couple planned an elaborate October wedding. Instead, the ceremony took place just one month later in the Frateses' home in Springfield, Missouri, at ten o'clock at night. It was attended only by family, and the couple left immediately after their vows, in a private railroad car bound for Clarion, Pennsylvania. It was an unusual beginning, but indicative of the far-from-ordinary life that they would build together.

While mysterious in the newspaper accounts of the day, the wedding's "shotgun" appearance was explained by the fact that Slick's older brother was seriously ill. It would not have been proper for Berenice—a single woman—to accompany her fiancé to Clarion in 1915, so they married immediately and rushed to Pennsylvania that very night. Jesse recovered from that illness, but his health continued to decline; he died four years later, sending Tom Slick into another of the many depressions that plagued him over the next decade.

But on May 6, 1916, Slick was far from depressed. In Oklahoma, a new gusher came in, providing him with a now-familiar rush of success. Oil production in the Cushing Field had surpassed 300,000 barrels a day, with most of it sent to fuel US forces fighting in World War I. But the birth of a baby son in Pennsylvania produced a completely new experience. For the rest of his life, Slick would feel the pull between work and family, the joys of home life competing with his gypsy spirit, and Clarion would remain the place where both feelings coexisted in temporary harmony.

Berenice Frates Slick never much liked the Clarion house, and she ruefully remembered arriving in her husband's hometown on their wedding night, where he presented her with a special gift—"a dark, three-story monstrosity" of a Victorian house across the street from her mother-in-law's home, already completely furnished for the new bride. According to family stories, it took some getting used to. She wrote home with dismay that the house was terrible, and to make matters worse, "they put sugar in their cornbread up here."

Thomas Bernard Slick was born during that first summer in Clarion, named for his father, with his middle name chosen in honor of the man who introduced his parents. (Later, after a falling-out with Bernard Jones, the Slicks would change their son's name to Thomas Baker Slick Jr.) The next year, their daughter, Betty, was born, followed by Earl in 1923. All three children resembled their father, fair-haired and blue eyed, and somewhat slight in stature.

Slick Sr. would slip in and out of the scene, moving between Pennsylvania, Oklahoma, Mississippi, Kansas, Texas, and wherever else a deal might beckon. When he was with his family, he captivated them with his unique combination of tenderness and adventuresome spirit. When his little daughter, Betty, had nightmares, he sat by her bed and listened,

Tom Slick's wedding present to his bride was a three-story Victorian house, already furnished, in Clarion, Pennsylvania. *Betty Slick Moorman Photo Collection*

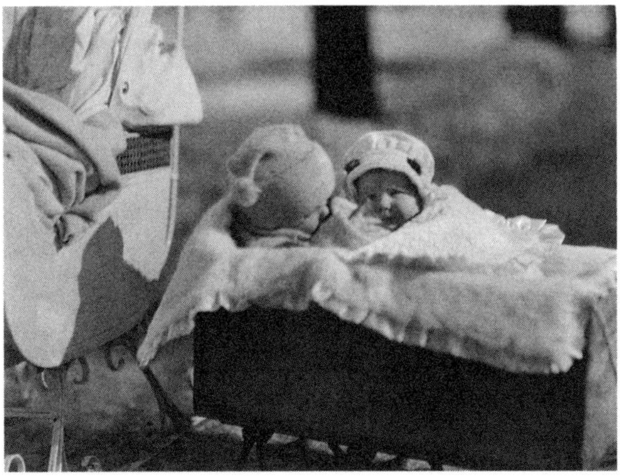

Tom and Betty Slick were just one year apart in age; they would grow up to be best friends. *Betty Slick Moorman Photo Collection*

soothing her back to sleep. They had long conversations about the ghost that Betty heard bumping around the music room on the third floor, deciding it was "friendly." His sister Flo lived nearby with her husband, Charles Urschel, and their son, Charles Jr., who often joined his cousins in their search for the source of the mysterious attic noises and other adventures.

Slick shared stories with young Tom about the mysterious "L'Homme des Nieges" ("Man of the Snows"), who according to French explorers inhabited a place described as the "Roof of the World." When King Tutankhamen's sarcophagus was discovered in Egypt in early 1923, newspaper stories about the adventure captured Tom's active imagination, and he told his parents and sister that he would make equally exciting discoveries someday.

Slick's children adored their father, accepting the fact that he was a complicated man. Their mother was more remote, and despite her day-to-day presence in their lives, their relationship with her was rather for-

left–right: Betty Slick, Charles Urschel Jr., and Tom Slick were cousins and playmates in Clarion, Pennsylvania. *Betty Slick Moorman Photo Collection*

mal. Decades later, Betty Slick Moorman, a mother and grandmother herself, would explain, "she was scary," hinting that her father's constant travels were partially to escape Berenice's critical personality.

But summers were special family times in Clarion, and during one of his sojourns there, Slick Sr. took young Tom to explore the nearby woods. They discovered a fallen log across the river that would eventually become a favorite hideout for Tom and Betty on Sundays, when they played hooky from church and invented all sorts of games and adventures in a natural playground of trees and twisted trails.

But the first time that father and son visited the log, Tom was only five or six years old; he was frightened when his father urged him to cross the river by walking on the log. The older Slick recognized an opportunity to convey one of his famous "lessons in life."

"A coward dies a thousand deaths; a brave man only one," Slick quoted to his son in their spectacular forest classroom. The little boy seemed to grasp the lesson's meaning; he walked the log. Later in life, he met adult challenges the same way, often remembering his father's words in the Pennsylvania woods and quoting them whenever he crossed new territory where no one else had dared to go.

The woods weren't the only exciting places to visit. Grandmother Mary Slick's house was "huge and haunted, with a hidden tunnel to China in the backyard," according to Betty Slick Moorman, who played there with her older brother. Two maiden aunts named Frances and Margaret lived in another house in the compound, and they provided even more mystery.

"Aunt Frances never spoke a word to us. She was very, very shy, and she never spoke to anyone except her two sisters—my grandmother (Mary Slick) and my Aunt Margaret. But she was so tender and sweet. She would play dolls with me, but she wouldn't speak. She would take Tom and me into her lap and then say to Aunt Margaret, 'Do you think the children would like to stay for dinner?'"

Eccentric Aunt Frances, and her sisters Margaret and Mary, were first-generation Pennsylvanians. Their father, Thomas Baker, came to the United States from England in 1843 or 1844, settling with his wife in Shippenville, just five miles west of Clarion. The couple established the first commercial flour mill in the area; they prospered and raised eight children in their large home in the center of town.

Mary Baker was the youngest daughter, and during her teens, she met an adventuresome young man named Johnson Slick, whose genealogy is a little sketchier than Mary's. The son of a tailor from Clarion, he left home at fourteen to work in the Bakers' flour mill, where he met Mary. Ten years later, in 1880, they married; they had three children—Jesse (1880), Tom (1883), and Flored (1888).

By 1892, Johnson and Mary Slick were living in Clarion, operating their own flour mill. But when the first oil in the United States was discovered in western Pennsylvania, Slick decided to enter this new business as a contract driller. His son, who joined him in the oil fields when he was sixteen years old, would become the largest independent oil operator in the world a little more than a decade later.

This family history, and the colorful characters who inhabit it, supplied fascinating subjects for discussion as young Tom and Betty Slick sat on the banks of the Clarion River, while their mother took care of baby Earl at home. Stories about how their great-grandparents met behind a flour sack, the secret reason for their shy Aunt Frances's silence, why their other grandmother "Muddy" (Lula Montez Buck Frates) whistled before she fainted, had all sorts of plots and subplots with mysterious aspects to explore.

Eight-year-old Betty, still convinced that at least one of her long-dead ancestors made ghostly visits to the Slick family, playing the piano in the music room, romping in the children's playroom on the third floor, enjoying all the secret nooks of the huge old house that might have been haunted, extracted a promise from her brother to investigate. More than thirty years later he kept that promise.

It's clear from family stories that the King of the Wildcatters encouraged his children's sense of adventure and love of wild landscapes. "Whenever our father came to Clarion to spend time with us in the summer, we always walked down to the river and the woods. He was so busy that he wasn't there much, but when he was, we had wonderful adventures," Betty Slick Moorman recalled more than seventy years later.

Slick never learned to balance work and leisure, although sometimes he made dramatic efforts to do so, recognizing that he had jeopardized his physical and emotional health by his frenetic work pace and lifestyle. His son would not make this same mistake; instead, he integrated work and leisure until they were one and the same.

But like his father, he was a searcher with a gypsy spirit, and both men paid a price for their frequent absences from home. Wives and children never quite understood.

Slick Sr.'s absences took him to Indian Territory in 1907, before he drilled his first big well in Oklahoma, when he still was known as "Dry Hole Tom." Years later, now nicknamed "Lucky Tom," he continued the friendships he had made in the Territory, always treating his Native American partners with a respect and fairness that were rare in those days. During his ascent from "Dry Hole Tom" to "Lucky Tom," Slick always discounted the importance of luck in the process of oil discovery; however, he did give some credit to intuition, experience, and knowledge.

"I know all of this country, every foot of it. As a leaser I drove and walked over all of it, studied, have learned to sort of sense, by intuition, where there ought to be oil," he explained when reporters asked how he'd become the world's largest independent oil operator in 1930, with net assets of $17 million.

In more homespun terms, he compared his so-called luck to a flock of ducks. "Did you ever shoot ducks?" he asked the newspapermen. "Lay hidden in a blind and call to the ducks as they fly over, and bring them within range? Same way with Luck. A whole lot of lucky chances may fly past unless you know how to call them, have a good gun, all loaded, and are able to shoot straight."

Not wanting miss any of the lucky chances flying over, Slick Sr. worked ferociously during the 1920s, branching out from his legendary oil successes to partner with his father-in-law, investing in plantations in Mississippi, building railroads, and establishing new townsites in Oklahoma named Slick and Nuyaka. And young Tom watched, and learned, and waited for his turn.

Like all games of chance, the stakes got higher as Slick's businesses expanded; eventually, his workaholic ways upped the ante to include his physical and mental health. Clarion always called him home, and quiet days with family were precious before he was again lured away. Throughout the 1920s, as his children, Tom, Betty, and Earl, explored the river and pretended that their woods were mythical kingdoms, Slick Sr. moved between the serenity of his birthplace and the much wilder world beyond. With new advances in technology and transportation,

FLOOD EDITION NO. 3

PIONEER OIL HERALD

VOLUME 1, NUMBER 10. PIONEER, EASTLAND COUNTY, TEXAS, TUESDAY, MAY 16, 1922. EVERYWHERE 5c

World's Biggest Shallow Oil Field

7500 Barrel Oil Gusher, Slick & Pennant Bryson No. 1 Blew in Roaring Out Bowels of Earth from 2457 ft. Sat. Night

THE WORLD'S BIGGEST HIGH GRAVITY GUSHER OIL WELL BLEW IN LAST SATURDAY AFTERNOON AT FIVE O'CLOCK WITH A ROAR; GUAGING 390 BBLS. OIL 43.7 PER CENT GRAVITY FOR ONE HOUR AND FIFTEEN MINUTES, OR 7488 BBLS FOR EACH 24 HOURS. IT IS OWNED BY T. B. SLICK AND THE PENNANT OIL AND GAS CO., WHO ARE MESSRS. O. V. GOLDRICK, W. C. BAKER AND M. M. HARRELL.

10,000,000 FEET WET GAS STAR NO. 1 NEAR CROSS CUT

WELL EXTENDS PIONEER OIL FIELD TEN MILES —IS TRAPPING SEVEAL BBLS. PURE GASOLINE

By the mid-1920s, Tom Slick Sr. had become the largest independent operator in the world. *Betty Slick Moorman Photo Collection*

the world was expanding. In 1926, National Broadcasting Company linked twenty-four radio stations to form the first "network," bringing together news and features from across the United States; the next year, aviator Charles Lindbergh made his historic solo flight across the Atlantic Ocean, flying from New York to Paris in thirty-three and a half hours; and by 1928, US oil production had more than doubled from what it was in 1920.

After nearly twenty years in the oil business, Slick was once again exhausted, and his family began to plead with him to take a vacation. In March 1929 he sold all his oil producing lands to the Prairie Oil and Gas Company for $40 million, announcing that he was going to "rest for a while." Newspaper and magazine articles described the details of the sale, estimating the oil fields he sold were yielding about 34,000 barrels a day at the time.

Slick kept part of the royalties on the production he sold to Prairie, and he still owned 525,000 acres of leased, undeveloped land. One observer remarked that the Slick deal was "like selling the milk but keeping the cows"; friends and competitors wagered that he would not be able to resist returning to the oil business. But Slick had promised his family a vacation, and with the huge cash transaction completed, he gave farewell bonuses of two million dollars each to his colleagues, E. E. Kirkpatrick of Tulsa, who ran the Slick Company's oil operations, and his brother-in-law, Charles Urschel, who had been a business associate for the past ten years.

Slick's health had been poor for several months, but the extraction of an infected tooth seemed to have improved it. The family arranged a trip to Texas, with plans to spend time on the Gulf Coast where Slick would fish and at his ranch south of San Antonio where he loved to hunt. Just the year before, he had hosted a large group of oilmen at a special dinner at the Tulsa Country Club, where he served nothing but wild game he had killed on his ranch. It was described in the local newspaper as "one of the most successful affairs ever staged on the midcontinent"; Slick planned to repeat it in the fall with a new supply of wild game.

As the family prepared to travel, Slick told friends, "I am a mass of nerves and a bundle of nerve. I need a good rest, maybe for a year or so." But just one month later he was back. In April he filed papers in New York City to incorporate the Tom Slick Oil Company in Oklahoma. Executives of the new company were a familiar cast of characters: Slick was president, Urschel was first vice president, Kirkpatrick was second vice president, and J. H. Grand was secretary.

The new company began operating out of Tulsa, with an office and activities in Texas as well. The pace was soon intense; Slick was again in the oil game and back to his workaholic ways.

About a year later, he checked into Johns Hopkins Hospital in Baltimore after a breakdown, much like the one he had in 1915. He was diagnosed with thyroid disease, and he was so run down that a "rest cure" was prescribed on June 27, 1930, until he was strong enough for surgery to remove a goiter.

On August 5, the operation was performed, and surgeons thought it had been successful. As he recovered, the Slick family began to plan

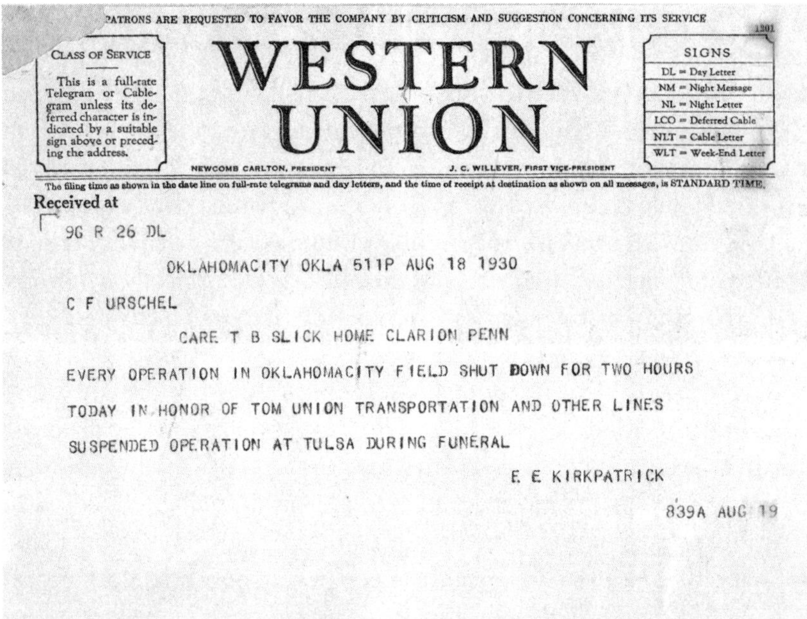

PATRONS ARE REQUESTED TO FAVOR THE COMPANY BY CRITICISM AND SUGGESTION CONCERNING ITS SERVICE

CLASS OF SERVICE

This is a full-rate Telegram or Cablegram unless its deferred character is indicated by a suitable sign above or preceding the address.

WESTERN UNION

NEWCOMB CARLTON, PRESIDENT J. C. WILLEVER, FIRST VICE-PRESIDENT

SIGNS

DL = Day Letter
NM = Night Message
NL = Night Letter
LCO = Deferred Cable
NLT = Cable Letter
WLT = Week-End Letter

The filing time as shown in the date line on full-rate telegrams and day letters, and the time of receipt at destination as shown on all messages, is STANDARD TIME.

Received at

9G R 26 DL

OKLAHOMACITY OKLA 511P AUG 18 1930

C F URSCHEL

CARE T B SLICK HOME CLARION PENN

EVERY OPERATION IN OKLAHOMACITY FIELD SHUT DOWN FOR TWO HOURS TODAY IN HONOR OF TOM UNION TRANSPORTATION AND OTHER LINES SUSPENDED OPERATION AT TULSA DURING FUNERAL

E E KIRKPATRICK

839A AUG 19

When Tom Slick Sr. died in 1930, Oklahoma oil rigs shut down in his honor.
Betty Slick Moorman Photo Collection

a two-week vacation in Canada. Fourteen-year-old Tom was looking forward to fishing with his father, and to seeing his grandmother, who would travel from Clarion to join them.

But the following Friday Slick had a stroke while sitting on the porch at the hospital. It was followed by a second one, and a cerebral hemorrhage. The King of the Wildcatters was dead at forty-six.

A special train took his body to his hometown of Clarion for burial, and family, friends, and fellow oilmen from the Southwest traveled east to pay tribute to him. Hundreds of wires and cables of condolence arrived, indications of the respect and high esteem felt for the oilman, not only by leaders in the industry but also by the men in the field. At 3:30 p.m. on Saturday, as last rites were performed in Clarion, oil wells in Oklahoma and Texas stopped drilling for the rest of the day in honor of the famous wildcatter.

The oil industry had lost a leader; fourteen-year-old Tom had lost a father. As the new decade began, the boy felt the huge void created by his father's death. He offered comfort to his little sister and brother, and he felt the responsibility of the Tom Slick legacy. He knew that his father had forged that legacy in the oil fields and remembered his claim that he was "born among the oil derricks. The first sniff of air I ever breathed into my nostrils was laden with the odor of oil."

That odor would permeate the lives of the Slicks for generations—it had lured a father to remote oil rigs far from home, would later attract a kidnapper with its scent in the 1930s, and eventually would provide the incredible wealth that would open doors to all sorts of strange new worlds for young Tom Slick.

Wife and Three Children Given $75,000,000 Estate

★ ★ ★ ★ ★ ★ ★

Mrs. Tom Slick Becomes Oil Operator Under Terms Of "Wildcat King's" Will

Mrs. Tom B. Slick

Vast Holdings of "King Of Wildcatters" Will Be Administered by Mrs. Slick, Urschel and Seeligson As Trustees.

Income For Youngsters

Their Share Will Be Made Available in Three Instalments, Paid at Ages of 30, 45 and 55 Years.

CLARION, Pa., Aug. 27.—(AP)—An estate estimated at between $75,000,000 and $100,000,000 was disposed of in the will of Thomas B. Slick, independent oil operator, which was probated Wednesday. Slick was reputed to be the wealthiest independent oil operator in the world.

The vast fortune was left to the widow, three children and Slick's mother. While no estimate of the value of the estate was contained in the will, Slick's attorney and others who were closely associated with him provided an estimate of its worth.

The oil operator's mother, Mrs. Mary Slick of Clarion, was provided with $5,000 a year as long as she lives.

The widow, Bernice Slick of Oklahoma City, Okla., was left one third of the remainder of the estate and the remaining two thirds was left in trust to the three children.

The Slick home in Oklahoma City also was left to the widow.

The three children are Thomas B. jr, Betty and Earl. At the age of 18 the daughter will receive $5,000 a year until she is 30 years old. The sons, at the age of 21, will receive $10,000 a year each until they are 30. At the age of 30, each child will get one third of the amount due him or her; at the age of 45 another one third, and the last one third at the age of 55.

The widow, Charles F. Urschel and Arthur A. Seeligson, were appointed trustees and empowered to either sell or carry on the Slick oil operations and other business.

The will was drawn in Baltimo[illegible] Md., on July 15, 1930, while Slick [illegible] a patient at Johns Hopkins hosp[illegible]. Slick died at the hospital and the body was buried here, his boyhood home.

The oil operator was known as "the king of the wildcatters." He engaged

The King of the Wildcatter's estate was huge in 1930, and publicity surrounded it for months. *Betty Slick Moorman Photo Collection*

3

Alone

The 1930s

Oklahoma City was booming in the 1930s, and oil was the impetus. The Slick mansion on Eighteenth Street was a beautiful three-story Victorian, with a large wooden porch that circled the house. Moving between their houses in Pennsylvania and Oklahoma had become a way of life for the family, but now that she was widowed, Berenice Slick left the house she never liked in Clarion and moved to Oklahoma, joining her parents and several siblings. The summer of 1930 was a sad one; Tom and Betty particularly missed their father.

In a gesture of family support that was quite common in those days, Berenice's youngest sister, Ramona, her husband Arthur Seeligson, and their two sons, Arthur and Frates, moved in with the Slicks. Seeligson was an attorney and the executor of his brother-in-law's huge estate. He was also a kind, family-oriented man who provided fatherly support to the young Slick children. Tom, Betty, and Earl were delighted to have their cousins living in the house, and a summer trip to the beaches of California provided activities and laughter that eased a difficult time.

In 1931 the family consolidated still more. Charles Urschel, Slick's oil partner and brother-in-law, had become a widower when Flored died the year before her brother's death. The families always had been close during the early wildcatting days and through their rise to wealth and social status. After a brief courtship, Urschel married Berenice, and his son, Charles Jr., was now a stepbrother to Tom, Betty, and Earl, in addition to a first cousin.

Tom Slick was fourteen when his father died; the loss instilled in him feelings of responsibility for others that would last a lifetime. *Margaret Urschel Photo Collection*

The death of his father when Tom was fourteen years old would of course shape the boy's personality in profound ways. Despite his mother's remarriage, and the eventual success and security that his sister and brother would establish, he assumed the role of the "man of the family" and would feel a sense of responsibility—extending beyond his immediate family—for the rest of his life.

Betty and Earl both described their big brother as a gentle leader, someone they turned to for advice and support, and family and friends that came along later reported the same.

After completing ninth grade at Classen High School in Oklahoma City, Tom was enrolled at Phillips Exeter Academy, a respected preparatory school in New Hampshire that listed some very impressive alumni, including one US president (Ulysses S. Grant), three governors of New Hampshire, several secretaries of state, and many well-known scholars, authors, and college presidents. As early as 1908, the school could claim that graduates of Exeter had held all the major offices in Presidential cabinets.

Education in the United States was undergoing real change. While many of the "self-made" men of the early 1900s had made fortunes without much schooling, they wanted more for their own children. Slick Sr. had not even completed high school when he left home for the oil fields; much more was expected of his son Tom.

But according to his Aunt Ramona, his earliest days at Exeter "were no indication of his future achievements." She remembered traveling by train from Oklahoma City to New Hampshire, with her husband Arthur, sister Berenice, and Tom. They stopped briefly in New York City to see the newly completed Empire State Building, the tallest building in the world.

"Tom was an overweight, reluctant and unhappy teenager," she recalled. "And unfortunately, added to this, his face had a purplish ointment look, due to prolonged treatment for impetigo. Only someone who truly loved him would have had the heart and nerve to claim him as family before any Admissions Board, much less Exeter's."

During his first few weeks at the school, he constantly wrote letters to his mother, begging to come home. He was lonely, and academically he was not prepared for the elite prep school. In his opinion, everything was wrong with the Northeast, and Tom began to think of creative reasons why he needed to return to Oklahoma.

In a letter written in November he explained that everyone at school had a "horrible Eastern disease called 'pink eye'; only leaving Exeter would prevent his infection. The Urschels were unyielding, and Tom began to consider his other options, a characteristic for coping with challenges that he would refine and call on throughout his life. "You've got to adapt or you'll become extinct like the dinosaurs," he would tell his children, partners, and anyone else who would listen to his words of wisdom in future years.

He discovered a beautiful room on the fifth floor of Phillips Hall, with Chinese green walls and rows of dark brown books, circles of comfortable chairs, and a mahogany rostrum for public speaking. He learned that Exeter was famous for the ancient avocation of its young men—public speaking—and debate became a new tool for making friends and "convincing enemies." Squash also captured his enthusiasm, and he became the team captain by his senior year.

Exeter also offered state-of-the-art science laboratories, where Tom experimented in chemistry and physics. The school's Preventive Medicine Infirmary was ahead of its time, reporting that the use of ultraviolet treatments had built up resistance to the common cold among the student body. The teenager was fascinated, and by the time he went home for summer vacation, he was an Exeter man.

In the summer of 1933, the Slick/Urschel family was in Oklahoma City. Shortly after 11:00 p.m. on Saturday, July 22, two men with guns walked onto the front porch of the Urschel mansion, where a friendly bridge game was underway in the cool evening air. Berenice screamed.

"Stop that or we'll blow your heads off," George "Machine Gun" Kelly is reported to have said. He asked that Charles Urschel identify himself, and when neither man replied, Kelly and his accomplice, Albert Bates, loaded both Urschel and his bridge partner, Walter Jarrett, into a waiting automobile.

Once the kidnappers were outside of town, they searched their captives' wallets and quickly discovered which man was Urschel. Taking the fifty-one dollars in Jarrett's wallet, they pushed him out of the car onto a country road and drove away.

Meanwhile, the wives had run inside and telephoned the police. Just a few years before, the Lindbergh kidnapping had horrified the world; the women were terrified. The police immediately put Berenice in touch with J. Edgar Hoover, director of the FBI in Washington, DC, and within an hour, the house was filled with local police, federal agents, and the press.

Soon, everyone had a good description of the kidnappers from both wives, and from Jarrett, who arrived back at the house around 3:00 a.m., tired and bedraggled. There is some evidence that the kidnappers had originally planned to take fifteen-year-old Betty, based mainly on the fact that the teenager reported seeing the two men following her on two occasions earlier that week.

Urschel and newly elected president Franklin D. Roosevelt were friends, and Hoover recognized that it would be essential to crack the case fast. While lawmen and federal agents in Oklahoma City began an intense investigation, Kelly arrived at his gang's hideout near Paradise, Texas, a five-hundred-acre ranch that belonged to his wife's stepfather, "Boss Shannon." Urschel was blindfolded and handcuffed and taken to a shack adjacent to the ranch house, where he remained captive for nine days. Kelly read him the newspaper headlines about the kidnapping and informed Urschel that there were too many federal agents working on the case to safely send a ransom demand to his wife in Oklahoma City.

A creative thinker, Urschel suggested using a family friend in Tulsa, John Catlett, as a messenger, and the kidnappers gave him a pencil and paper and instructed him to write two notes—one to Catlett and one to his wife.

As the ransom note made its way toward Oklahoma, Urschel was learning as much about his surroundings as he could; despite his blindfold, he collected valuable information that would later help the FBI. He noticed that he could hear airplanes fly overhead twice a day, and when they did, he asked his captors what time it was. Later he would tell federal agents that at 9:45 every morning, and at 5:45 in the afternoon, planes flew over the hideout. Since there were far fewer flights in 1933, this clue would be of major assistance in pinpointing the location of the hideout. He also had as many conversations as possible with everyone at the Shannon ranch, learning that Bates did not think much of the famous bank-robbing couple Bonnie and Clyde, and carefully remembering a new name—Harvey Bailey—when he arrived at the ranch with a wounded leg. He left his fingerprints on everything he could touch, hoping they would someday provide evidence that he was there.

On July 26, a messenger delivered a large envelope to the Catlett home in Tulsa. It contained the two notes from Urschel and a ransom letter from the kidnappers to E. E. Kirkpatrick, an oilman/rancher and newspaperman, who was also a close family friend. It demanded $200,000 and gave instructions for delivering it.

The kidnappers wanted "genuine used federal reserve currency," and they outlined exactly what the Urschel family should do next. Following directions, they ran this ad in the *Daily Oklahoman* newspaper:

> *FOR SALE—160 acres land, good five-room house, deep well.*
> *Also cows, tools, tractors, corn and hay. $3,750.00*
> *For quick sales TERMS*

Two days after the ad ran, an airmail letter arrived at the newspaper, postmarked Joplin, Missouri, instructing Kirkpatrick to board the train for Kansas City on the night of July 29. He was told to travel alone, and to watch for two successive bonfires on the right side of the track. Once he saw the second fire, he was to throw the ransom money off the train. The kidnappers said Urschel would be permitted to return home once they received the money. They also provided a contingency plan in case something went wrong, telling Kirkpatrick to check into the Muehlebach Hotel in Kansas City under the name of E. E. Kincaid and wait for further instructions.

The Urschel family withdrew the money from First National Bank of Oklahoma City, and the FBI recorded all the serial numbers. Catlett boarded the train with Kirkpatrick, which he was not supposed to do. Whether the kidnappers knew this, or as some sources report, Kelly had car trouble and could not set the bonfires in time, Kirkpatrick and Catlett had an uneventful train ride to Kansas City, and they checked into the hotel late Saturday night.

The following morning, a telegram arrived, promising to contact Kirkpatrick later that day. The telephone in Kirkpatrick's room rang at 5:45 p.m. on Sunday, July 30. A man's voice directed him to take a taxi to the LaSalle Hotel and walk west, with the suitcase containing the money in his right hand. This time the instructions were followed almost exactly—Kirkpatrick did tuck a gun in his belt—and Kelly met him on Linwood Boulevard, saying, "I'll take that grip," assuring him that Urschel would be home in twelve hours.

Family and friends remembered the incredible tension of the entire week, and the excruciating wait on July 31. While Kelly and Bates divided the money at the Shannon ranch, they debated about what to do with Urschel, finally opting not to kill him. After allowing him to bathe and shave, they drove through rain to the outskirts of Oklahoma City, warning him not to talk about his kidnapping. They told him to walk a few miles to a gas station where he could call for help, and they gave him ten dollars for cab fare. Around nine o'clock that night, Charles Urschel arrived at his home. Berenice and the children were waiting, along with the FBI. The manhunt for the kidnappers would begin the next day.

With the information that Urschel supplied about airplanes flying over at specific times, the FBI was able to pinpoint the Shannon ranch, and on August 12, a raiding party of federal agents, local police, and Urschel arrived. Harvey Bailey was there, with several others who had just robbed the Peoples National Bank in Kingfisher, Oklahoma. The gang was arrested, along with Boss Shannon, his wife Ora, and son and daughter-in-law. Some of the ransom money and machine guns belonging to Kelly were collected as evidence.

That same day, Bates was arrested in Denver, Colorado, when he tried to use stolen money orders from an earlier bank robbery. George and Kathryn Kelly were on the run, growing more nervous as they learned of each new arrest.

Amid the summer drama, Tom was packing for his senior year at Exeter. Before leaving for New Hampshire, he saw an odd-looking envelope arrive at home, with a note threatening the entire Slick/Urschel family if the Shannons were convicted. The chase for the Kellys intensified, taking some bizarre twists, and on September 26, just as Tom was getting settled at school, the gangsters were arrested in Memphis. They were extradited to Oklahoma and tried for kidnapping Charles Urschel.

The trial was well documented since it set many new legal precedents for kidnapping cases. It was dramatic and colorful; the jury deliberated for less than an hour before finding George and Kathryn Kelly guilty on October 12, 1933. They were sentenced to life in prison, and Kathryn Kelly would say on many future occasions that she wished they had killed Charles Urschel.

The Urschels and members of their extended family that included siblings, aunts, uncles, and cousins, never forgot the terror of the kidnapping. As a result, Tom, Betty, Earl, and stepbrother Charles developed a strong aversion to publicity, and their families worried for several generations that somehow the Kellys and their gang would seek retribution.

Years after the kidnapping, Urschel and Kelly corresponded, and eventually Urschel visited his kidnapper at Alcatraz prison, reporting that he was "a changed man." Kelly died in prison in 1954, one day after his fifty-fourth birthday. Kathryn Kelly was paroled in 1958, after twenty-five years in prison, and died in Tulsa, Oklahoma, at the age of eighty-one.

But as Tom prepared to graduate from Exeter in the spring of 1934, the Kellys were in prison and no longer a threat to his family. He saw an exciting future taking shape, starting with a wonderful graduation present from his grandfather, Joseph Frates.

Frates held a special place in Tom's heart. The boy resonated with his adventuresome spirit, and his love of hunting, fishing, and traveling reminded him of the father he missed. When the school term ended in 1934, Tom and Betty Slick, and their stepbrother Charles Urschel, all completed the year with honors. Berenice wanted to show her appreciation and asked her children what they would like. Tom suggested a trip to some foreign country, accompanied by his grandfather. Charles was quick to sign on to the idea.

"I was delighted, and the trip was planned," Frates wrote in his diary. "Betty, of course, was included in the award; but she wanted something else." Charles was attending Culver Military Academy in Indiana, Tom was at Phillips Exeter Academy in New Hampshire, and Betty was at the Miss Madeira School in Virginia. The extended family attended the end of term at all the schools and completed the circuit at Exeter Academy on June 23, 1934.

On June 30, Joseph Frates boarded the SS *Monarch of Bermuda* from the docks of Lower Manhattan, with his grandsons Tom Slick Jr. and

Charles Urschel Jr. The send-off party consisted of Charles and Berenice Urschel, Arthur and Ramona Seeligson, Mrs. Joseph Frates, Betty and Earl Slick, Earl Frates, and two schoolmates of Tom Slick's—Fred Kemple and Kemp Gugglemann. Powerful tugboats pulled the ship out of the harbor, to shouts of "bon voyage."

Tom Slick's first ocean voyage, accompanied by his grandfather and stepbrother, was destined for Rio de Janeiro, Brazil. Bermuda, Cuba, Panama, Colombia, Ecuador, Peru, the Guano Islands, Chile, Bolivia, and Argentina were on the itinerary prior to Brazil; the return journey included traveling over the mouth of the Amazon, past Cayenne and Devils Island, French Guiana, Venezuela, Trinidad, the Virgin Islands, Puerto Rico, Santo Domingo, and Haiti, finally reaching Miami, Florida, and returning to Oklahoma City, Oklahoma.

"We traveled by steam railroads, cog railroads, steamships, sea planes, and automobiles," Joseph Frates wrote in an account he circulated to family members as a Christmas present in December 1934. "We saw a great many interesting places, natives of many different countries and languages, came in contact with some high class people, well informed on conditions throughout the world, visited one country where war was, and still is, in progress with an adjoining nation, another where civil war was in progress. We also passed through the greatest inland waterway in the world—the Panama Canal."

Diaries record Tom's enthusiasm for everything new and exotic, and his grandfather chuckled when he ordered every unusual item on each foreign menu, sometimes "paying for it" the next day.

"While we were on this trip, Tom was continually complaining about the food; there were too many American dishes; he wanted native foods, something he couldn't get at home," Frates wrote in his diary. In Rio de Janeiro, friends who were part of Standard Oil's South American enterprise entertained the travelers. Aware of young Tom's desire to sample native foods, they went to a restaurant that served a local specialty—an octopus dinner.

"When the dish was served, I took a small bite and I want to tell you right now it was terrible," Frates wrote, "but of course I said nothing. Looking around the table, I noticed that our hosts were eating sparingly. I am not sure how the fish affected Tom; however, I do know that he was looking for pills that evening, and to the best of my knowledge, there was no further complaint from Tom about American food during the rest of the tour."

Tom and Charles explored every corner of the elegant ship, befriended the crew, and were sought after by young ladies on the dance floor. During the last days in Rio, Tom bought a monkey as a present for his younger brother Earl. Nicknamed "Johnny," the pet was mischievous, and according to Frates, "one evening Johnny got loose, and what he did to things in Charles' and Tom's rooms was plenty; the rooms looked like a cyclone had struck them."

The travelers departed the elegant Hotel Gloria, Oklahoma bound, with Johnny in tow. A twin-engine, twenty-two-passenger seaplane operated by Pan American Airlines flew them to Miami, and they completed their journey by train.

When Tom returned to Oklahoma to pack for his freshman year at Yale University in New Haven, Connecticut, he had no doubt that he would travel to every corner of the world in the years ahead.

Slick was a member of Pierson College at Yale, and he lived alone at 74 Wall Street. By the end of the year, he was second in his class, pursuing a premedical course of study. He entered athletics with enthusiasm, with great success on the squash and crew teams, and his love of public speaking, acquired at Exeter, eventually earned him a "high oration appointment" during his junior year.

Albert Einstein said that imagination was more important than knowledge, but Slick believed both were essential. His outstanding record at Yale, culminating in Phi Beta Kappa, reflected his love of knowledge, and his sister Betty reported that "he had enough imagination for everyone."

Only a year apart in school, they often spent college weekends together. Betty was attending Vassar College nearby, and she introduced her handsome brother to several attractive girls. She was already in love with Lew Moorman, whom she would marry in 1938, before finishing college, and wonderful memories of double dating still were vibrant many decades later.

According to Betty, Tom's pursuits in college—both scientific and romantic—were legendary. "His interest at the time was in the possibility of cross breeding various and sundry animal types and species.

Tom Slick attended Yale University in 1934 and graduated Phi Beta Kappa with a degree in science in 1938. *Betty Slick Moorman Photo Collection*

Betty Slick attended Vassar College and married her sweetheart Lewis Moorman in 1938. *Betty Slick Moorman Photo Collection*

He carried in his wallet a list of possibilities, such as camel and horse, cow and buffalo—which actually was done later, but not by Tom—culminating in ape and man. His girlfriends viewed this list with mixed emotions and did not always find it amusing.

"The only concrete accomplishment to result from this interest was his purchase of the 'Hoat,' more affectionately known as Sweet William. Tom had read in *Ripley's Believe It or Not* about a farmer who had raised an animal that was presumably a cross between a hog and a goat. He sought out the farmer and purchased the 'Hoat,' which he kept at our mother's farm near Oklahoma City.

"Sweet William was a disreputable looking, deformed creature—high in the front, low in the back, with scraggly goat hair and an equally scraggly hog's tail. The goat tendency to be a horned animal centered in his hooves, which grew outward and upward, forming almost a complete circle. It was as if he walked on rockers.

"Tom exposed Sweet William to both goats and hogs for breeding purposes, but nothing happened. He probably was sexless, or so unattractive that no self-respecting goat or hog would have anything to do with him. Sweet William died a natural but early death. Luckily for all of us, Tom's scientific interests began to find expression in other ways."

Fascination with the Loch Ness Monster inspired Slick to organize a trip to Scotland with several fraternity brothers during the summer of 1937. The young men explored the local pubs, played golf on the famous course at Saint Andrews, but did not see anything of interest emerge from the famous lake. Slick later told his mother that he had proven something important on his expedition, "that science and fun can coexist."

By his senior year at Yale, he was seriously dating Betty Lewis, a Vassar classmate of his sister's, from Dallas, Texas. The romance was volatile, since both were extremely intelligent and opinionated. "If Tom said something was blue, Betty said it was red," friends would say years later.

The summer of 1938 was a happy one for the Slick/Urschel family. Tom graduated from Yale, Phi Beta Kappa, and Betty Slick married her beloved Lew Moorman. Her spectacular wedding cake arrived from New York City by train. Later that fall, their younger cousin Arthur Seeligson Jr. followed a new family tradition and enrolled at Yale.

Tom accompanied young Arthur to New Haven, and many decades later Seeligson remembered it as a very special time. "Tom was like a big brother to me, and it meant so much to have him there. And we shared a very unexpected adventure that fall, one I'll never forget."

The "adventure" was the famous Great Hurricane of New England, one of the most powerful and destructive storms ever to strike the East Coast. The hurricane made a twelve-day journey across the Atlantic and up the Eastern Seaboard before crashing ashore in Long Island on September 21. The eye was in New Haven, and so were the two cousins.

Winds of more than 120 miles an hour and storm tides reaching eighteen feet ripped across Connecticut. Downtown Providence, Rhode Island, was submerged under twenty feet of water. New England was devastated. More than 15,000 homes were damaged, and 8,900 were totally destroyed. More than 6,000 boats were either destroyed or badly damaged; 564 people died, and more than 1,700 were injured. Downed power lines touched off fires across Connecticut, and flooding washed away roadways and sections of the New York, New Haven, and Hartford railroad lines.

Seeligson remembered how glad he was that his older cousin was with him, how they listened together to the howling winds and talked about the incredible power of nature. Slick believed that storms, earthquakes, erupting volcanoes, and other natural forces were humbling experiences, reminders of a much larger universe where man was but "a tiny speck." He was certain that scientific research would help to harness some of these energy sources to improve human life, but he also was convinced that some phenomena would remain unexplained. He was comfortable with the dichotomy; he liked a world where logic, imagination, and mystery interacted in a special dance.

He brought the same forces to romantic life, and when he learned that his college sweetheart, Betty Lewis, was thinking of marrying a rival named John Young, Slick used his charm and proven oration skills to convince her not to. Energized by the challenge of competition, he sent flowers and thoughtful notes, and in January 1939, she accepted his proposal of marriage.

Betty came from a prominent Texas family and had been presented as a debutante in Dallas during the previous season. But with a degree in child psychology from Vassar, she intended to establish a career.

Earl Slick (*right*) hosted a western-themed bachelor party for his brother, to celebrate his upcoming marriage to Betty Lewis. *Betty Slick Moorman Photo Collection*

Her father, William Lewis, sensed that his intelligent, strong-minded daughter was not going to be happy with the equally intelligent, strong-minded Slick. But the young couple was determined, and their April wedding in Dallas was described in all the social columns as a splendid event.

But the marriage was stormy, and when Betty was pregnant with their first child, she moved back to Dallas from Oklahoma City, to be near her parents. The baby was born in January 1940, and William Lewis Slick was named for his maternal grandfather. His parents agreed to set aside their own bitter feelings about the short-lived marriage and subsequent divorce, and they tried to create a normal childhood for their son. Within a year Betty married John Young, and Slick experienced feelings of loss. But true to his nature, traits he had developed in childhood soon persevered. He did not fret for long. Instead, he accepted "outcome," as he liked to call his less than successful experiences, and he focused on the future. A new decade was beginning; and a world war would soon overshadow all else.

Slick's fiancé was glamorous and intelligent, with a degree from Vassar College. *Betty Slick Moorman Photo Collection*

With the outbreak of World War II in the Pacific, Slick volunteered for naval duty, but poor eyesight disqualified him. He served with the War Production Board in Chile and Washington, DC, until 1943, when restrictions eased and he was allowed to join the navy. *Courtesy of the Mind Science Foundation*

4

Dreaming Big

The 1940s

In June 1940, German planes bombed Paris, and five years later the Americans dropped the first atomic bomb on Hiroshima, Japan, killing more than 50,000 people and bringing an end to World War II. Postwar inventions like the long-playing record in 1948 and instant cake mix in 1949 seemed to promote a new sense of well-being in the United States, while half a world away, the Communist People's Republic of China was established by Mao Zedong after a bitter civil war. Tom Slick was a young man in his twenties, patriotic about his country, fascinated by science, and already thinking about ways he might use his family wealth to make the world a better place.

During the 1930s and 1940s, many large oil family dynasties, including the Slick/Urschel family, moved their home bases from Oklahoma to Texas. The reason was twofold: many of their major wells were located there, and there was no state income tax. In 1939, Tom and Earl Slick established the Slick-Urschel Oil Company in partnership with their stepfather and stepbrother, Charles Urschel Sr. and Jr. in San Antonio.

About this time, the senior Urschels built a spectacular home in the city's most exclusive suburb, designed by John Staub, one of the leading architects of the times. Berenice's youngest sister, Ramona, and her husband, Arthur Seeligson, joined the migration to San Antonio and built another beautiful mansion just a block away, designed by the same architect.

Slick began buying land, including the old Cable Ranch, on the outskirts of San Antonio, Texas, dreaming of creating a "Science City." *Courtesy of Southwest Research Institute*

On June 18, 1940, twenty-four-year-old Tom Slick bought a 1,600-acre tract of land on the outskirts of his new hometown, and during the next twelve months, he acquired more land in the area, including the 5,000-acre Cable Ranch, to bring his longtime dream of establishing a "Science City" on the Texas prairie to reality. He named his new ranch ESSAR, chosen to reflect his interest in scientific research, an acronym of sorts, combining an "s" (ESS) for science and an "r" (AR) for research, and spelled ESSAR. Slick planned for the site to become a cluster of basic and applied research organizations, where information would be shared and discoveries made that could improve the world.

At the close of 1941, just one week after the Japanese bombed Pearl Harbor Slick founded the first of five scientific research institutes that he would establish during his lifetime. The Foundation of Applied Research (FAR) would change its name three times in the decades ahead—to Southwest Foundation for Research and Education (SFRE) in 1952, to Southwest Foundation for Biomedical Research (SFBM) in 1984, and finally to Texas Biomedical Research Institute in 2011.

When Slick established FAR, he was just twenty-five years old, and to ensure its financial future, he endowed it with 1,875 shares of the Slick-Urschel Oil Company, announcing to anyone who would listen that "science gives us a tool of unparalleled effectiveness by which we can improve the physical side of our lives; and since science recognizes no geographic boundaries, the lives of people all over the world."

He had developed a list of criteria for his diverse endeavors, and it was evolving into his formula for life. He followed the formula each time he founded a new nonprofit organization; when he developed innovations in agriculture, cattle breeding, oil exploration, and construction; and when he organized expeditions to the Himalayas in search of unexplained phenomena. He started with a mystery, an unanswered question, a point of wonder. He assembled a team of experts. He planned

Slick and attorney Leroy Denman drew up legal papers to establish the Foundation for Applied Research (FAR) in 1941. *Courtesy of Texas Biomedical Research Institute*

a "capture"—of information, new oil wells, or yetis, for example. He insisted on a thorough investigation held to rigorous standards. And finally, results were shared whatever the outcome of the endeavor.

"I don't believe in failure," Slick said often, "only outcome."

"When there is a disappointment of some sort, I never think it is the end of the story. I believe it's the beginning of something new, sometimes a great adventure." Whether a scientific experiment, a new business venture, or a promising personal relationship, Slick's endeavors were imbued with optimism. And whatever the result, he found new opportunity and moved forward.

With the outbreak of war in the Pacific, Slick volunteered for naval duty, but poor eyesight disqualified him. Determined to do his duty, he found another way to be involved. In 1941 he joined the War Production Board in Washington, DC, and was sent to Chile, attached to the American embassy in Santiago. His years in the military as a "Dollar a Year Man" are not well documented, and many official records still have not been declassified. But there have always been rumors within the family that he was involved in special intelligence projects during the early years of World War II.

In June 2017, Chile's PDI (*Policía de investigaciones*) declassified more than a thousand files on Nazi activities there during World War II, sharing dramatic evidence that one spy ring headquartered in the port of Valparaíso had plotted to destroy the Panama Canal and bomb various mining sites in northern Chile. The German broadcasting company, Deutsche Welle, broke the story internationally, providing new details about a clandestine world of Nazi intrigue eight decades ago.

Four years earlier, the NSA (National Security Administration) shared its information on the "Cryptologic Aspects of German Intelligence Activities in South America during World War II" (Document ID: 3525901), which highlighted the US War Department's efforts to curtail the German espionage networks in Latin America. According to the NSA report, as early as 1940, the Nazi network was active in Chile, led by Heinz Lange (code name JANSEN). Focusing mainly on the movements of maritime shipping, it regularly intercepted radio communications by the Chilean Navy about allied ship routes, sending along coded information to the Third Reich. The port of Valparaíso was described as a hotbed of spies in the 1940s, and *Compañía Trans-*

Slick sent his sister a postcard from Santiago, Chile, describing it as "a hotbed of spies" when he was sent there in 1941 by the War Production Board. *Betty Slick Moorman Collection*

portes Marítimos (COTRAS), formerly a branch of Norddeutscher Lloyd, was implicated in the network, along with several high-ranking officials in the Chilean Coast Guard.

Just a few days before Hector Espinoza, director of the PDI, held his press conference in Santiago, Argentina announced the discovery of a Nazi cache of seventy-five artifacts hidden behind a secret door in an art collector's home. A statue of a Reichsadler, the imperial eagle symbol of the Third Reich, silver cups engraved with swastikas, Nazi medals, a bust of Adolf Hitler, and even a medical device for measuring human head size, were among the items seized and sent to the Holocaust Museum in Buenos Aires—grim proof of the Nazi presence in South America that has been the focus of countless books and films over the decades.

The US State Department charged the War Department with operation of its counterintelligence operations in 1940; Tom Slick joined the War Department in 1941 and was assigned to Chile. The personal correspondence and diaries in this book begin a few years later; one can only imagine what his duties were as a "Dollar a Year Man" in Santiago. But NSA records indicate that as a result of information collected by the Americans and given to the Chilean government, many members of the Nazi spy networks were arrested in the fall of 1942. Others escaped and built a new network, known as the PQZ group, which was run by Bernardo Timmerman until his arrest in 1944, when many Germans in Chile escaped to Argentina.

By that time, Slick was back in the United States, commissioned as an ensign in the navy in 1943, when eyesight requirements were relaxed. In the summer of 1944, he took courses at Harvard Business School and MIT as a government war student and was assigned to bases in Hawaii (as a liaison officer with the Oahu Railroad), Saipan (as commanding officer of an oil tank farm), and finally to occupied Japan.

Not long after he returned to San Antonio from his war duty, he met beautiful eighteen-year-old Polly Nixon, the daughter of a prominent Texas medical family (also originally from Pennsylvania) that was part of his mother's and stepfather's circle of friends. Nearing thirty years old, Slick was attracted to her youthful enthusiasm and femininity, and she seemed determined to become the perfect wife.

But the Nixons had reservations. Slick was divorced, something that was less accepted in the 1940s, and he was twelve years older than their daughter. They faced a difficult dilemma, caught between their friendship with the Urschel family and their dreams that their daughter would attend college and eventually marry a man closer to her own age.

The young couple was not deterred; the potential of it all was irresistible. While Slick saw a beautiful young woman willing to be molded into an ideal wife; Polly saw a handsome, worldly man who could provide her with everything she wanted, the fairy-tale prince that all girls dream might come their way. So, despite some resistance from their parents, they became engaged and planned their own wedding in New York.

Family members from both sides traveled by train to New York, and the Pierre Hotel provided a sophisticated, elegant setting for the wedding entourage. Slick's stepbrother, Charles Urschel Jr., was best man, and the bride's twin sister, Patty Nixon, was maid of honor.

Riverside Church in Manhattan was not far from the Pierre, and the wedding party began to assemble for a stylish six o'clock ceremony on

Miss Polly Nixon To Wed T.B. Slick

Polly Nixon was young, beautiful, and accomplished; Slick hoped to mold her into the perfect wife. *Betty Slick Moorman Photo Collection/newspaper clipping from* San Antonio Express-News, *1946*

September 4, 1946. But when the bride's older brother, Jim Nixon, did not arrive at the appointed hour, tension mounted. Many years later, he described his late arrival, caused because his watch had stopped, and everyone's relief when the organ music began. Dr. Nixon walked his petite, dark-haired daughter down the aisle; her huge dark eyes, creamy complexion, and elaborate ivory lace veil were a lovely vision; everyone hoped the fairy tale was real.

After a month-long honeymoon in Europe, the newlyweds returned to Texas, settling at ESSAR Ranch, in the old Cable House, an eighteen-room mansion that had once been the winter home of railroad tycoon P. L. Cable.

Later that same year, Slick met Travis Richardson, a straight-talking wrangler who had worked in intelligence along the Mexican border during World War II. Fluent in Spanish, his assignment had been to work up and down the Rio Grande, investigating rumors that the Japanese were planning to invade the United States from Mexico.

Just as Slick Sr. presented his bride with a house as a wedding present, the newly restored Cable House was waiting for Polly Nixon Slick on the outskirts of San Antonio when she married Tom Slick. *Courtesy of Southwest Research Institute*

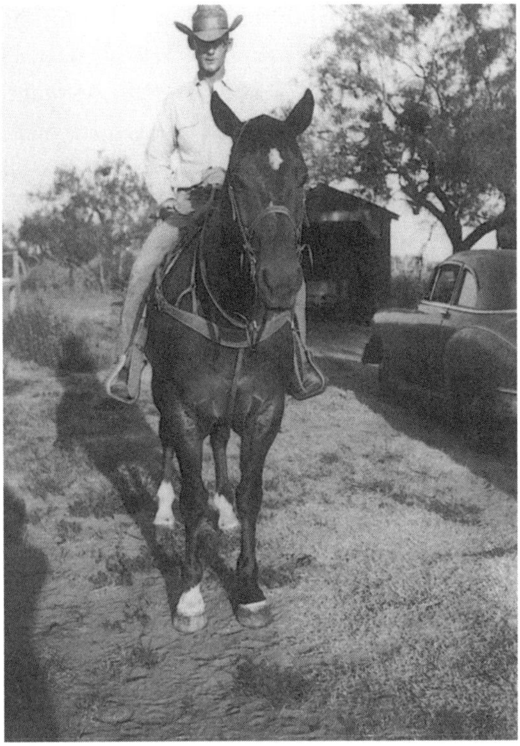

Travis Richardson was hired to oversee Slick's ranching operations; he helped develop the Brangus breed of cattle and participated in many Slick adventures. *Travis Richardson Photo Collection*

After the war, Richardson settled in Laredo, Texas, with his wife, Dorothy—a warmhearted woman who understood wanderlust and enjoyed it. Tom Slick met the Richardsons in 1946, liked them immediately, and asked them to move to ESSAR Ranch so that Travis could run its farming operation.

Nearly sixty years later, Richardson remembered his answer. "I've always enjoyed cattle more than anything," he told Slick wistfully. "I don't think I would like just farming." Slick considered the situation and did what would become a pattern in the years ahead. He liked the man, so he adjusted the job description to fit, agreeing to add cattle to existing farming operations.

Over the next few years, the family business consortium (Tom Slick and younger brother, Earl, brother-in-law Lew Moorman, and stepbrother Charles Urschel Jr.) bought a total of five ranches in South Texas and Oklahoma. The Richardsons lived on all of them, working with Slick on cattle deals, as promised. But Tom Slick could never just follow the established road, and soon, he and Travis were pushing the frontiers of the industry. By 1948, they developed a new breed—a cross between Brahman cattle from India, heat-tolerant and insect resistant, and the Scottish Angus, known for its superb meat quality.

"Tom was the first to develop the Brangus breed," Richardson said in 2001, sitting in the den of his ranch house in Pandora, Texas, filled with photos of adventures with Slick. "He had been to India and Scotland, and to many other places too, and he had studied the cattle everywhere he visited. He had the idea that breeding the hearty Brahman with the tastier Angus would produce better beef in cattle that could be heat and insect resistant . . . perfect for Texas."

There were some Brahman and Angus cattle closer to home, at the University of Louisiana, and Slick worked with scientists there to breed "Old Zero," a Brangus bull that eventually was taken to Clear Creek Ranch in Welch, Oklahoma (where the famous "hoat" once lived), and Travis Richardson was in charge of the entire Brangus operation by 1948.

In 1949, the International Brangus Breeders Association was established to oversee the careful registration of the new breed, defined strictly as three-eighths Brahman and five-eighths Angus, solid black, polled as to conformation and breed character.

Brahman cattle were first bred in the United States in 1885, from several breeds of cattle originating in India, and their thick skin made them extremely tolerant to heat and resistant to insects. In 1924 the American Brahman Breeders Association was formed to track and verify bloodlines of these hearty cattle that are known for their longevity, strong mothering instincts, outstanding milk production, and calving ease.

The Angus has long been considered the premier breed for beef quality, marbling at an earlier age than most breeds and reaching the desired grade and weight sooner. Early puberty, high fertility, and good milking ability contribute to the breed's success, especially when crossbred with the Brahman to produce the Brangus.

The Brangus was a beautiful specimen of breeding, according to Richardson. Because the two "parent breeds" were so outstanding in their respective strong traits, the new breed was the "best of both" according to Richardson. Other cattlemen and consumers agreed.

Brangus cattle—a cross between the Brahman and Angus breeds—were developed at the Foundation for Applied Research (FAR). *Travis Richardson Photo Collection*

"Tom was an amazing man," Richardson said as his wife served a gourmet lunch that included Brangus steaks in remote South Texas. "I would be in his office, and anybody in the world could call him and he would talk to them. His whole thrust was to help his fellow man, no matter what the venture.

"He didn't give a hoot in hell about making money on Brangus. He just hoped the breed would help people in poor, hot countries."

While the cattle at ESSAR, Clear Creek, and the other ranches flourished, Polly did not like the remote surroundings, and Slick's constant travels put a strain on the "fairy tale" relationship that his young wife had envisioned. In 1947, they moved into town. Their daughter, Patty, was born later that year, followed by son Tom in 1949 and Charles in 1951.

Slick organized his busy life to include short times spent with his family in San Antonio, always ensuring that they were filled with special fun, just as his father had done. But also like his father, long times were spent away from home—pursuing the new ideas that were pouring out of his research institutes, buying new ranchlands and experimenting with agribusiness, exploring new oil deals, and even looking to the sky for new projects.

Following World War II, the army had a surplus of Curtiss C-46 airplanes, which had functioned as the "workhorses" for moving equipment, supplies, and troops throughout Europe and the South Pacific. Earl Slick, who had flown in the US Air Force during the war, saw the chance for a bargain and bought nine of them for $247,000, against a cost of around $200,000 each to the army. An article in the July 1950 issue of *Fortune* magazine reported that "Earl Slick was so determined to get into the air freight business without any idea of what they were going to fly or where."

He suggested to his older brother that they create an airline devoted to cargo, rather than passengers, and Slick Airlines was established in 1946. Lew Moorman and Charles Urschel became partners as well, and for a while it led other carriers in market share. The planes flew everything from fresh cut flowers and okra to heavier loads like prize bulls and a six-ton bearing destined for US Steel in Pittsburgh. The organization employed mostly war veterans, and their dedication to the company made Slick Airlines unique. But complex struggles with passenger airlines including American, United, and TWA, which saw the freight airline as a potential competitor, and with the Civil Aviation Bureau of certification to operate as a common carrier were costly, and bitter battles over freight rates also affected the company's profits. Tom Slick and Charles Urschel had returned their focus to their oil business after a few years, while Earl Slick and Lew Moorman kept the dream alive.

By 1950, Slick Airlines had twenty-one planes hauling 2.5 million ton-miles over a route that included fifty-four cities in the United States, but it was suffering financially. In 1954 a merger with another freight carrier, the Flying Tiger Line, looked promising, and both companies were disappointed when the deal collapsed over branding issues.

For a period in the late 1950s the airline shut down its operations, and rumors persist that the planes were used for covert CIA (then INS) operations until the company resumed its flights in 1962. There is documentation, with portions classified, that a Slick Airways plane landed at an Arizona military base when the airline was supposedly suspended. In 1968 the company was acquired by Airlift International and used mainly to fly the wounded and dead back from a new war underway in

World's Biggest Air Freight Line

VOLUME UP 250%

Slick Sets the Pace

Air freight hauling is revolutionizing America's vast system of distributing her gigantic production of things people use.

And Slick Airways, Inc., whose home office is San Antonio, is leading the way.

Slick, the world's biggest air freight line, is hauling at a five per cent cheaper rate than in 1948, though costs are up 20 per cent. The answer: volume is up 250 per cent.

Even with this terrific increase in volume, Air Freight has just scratched the surface and its total tonnage is only a small portion of the tonnage of American railroads and truck lines.

David Stewart, vice-president and treasurer, said air freight carriers can hold their relative places in the field, increase their tonnage four times and still be but a "flea bite" on the other carriers' total tonnage.

He believes air freight is but an "extension of the market" for business that otherwise could not meet those "extended markets." As an example, he cited the Rio Grande Valley okra grower who is doing very well shipping okra to Detroit via Slick. Without air freight, he couldn't meet demands in that distant market.

The automotive industry has taken to air freight enthusiastically. Stewart, who was associated with General Motors for 17 years, said that when "the most cost-conscious industry in the world starts using air freight, it's time other businesses study it too."

A point Slick is proud of, too, is that there has been no government subsidy back of the line.

Slick is one of four certified lines that aren't subsidized. Other three are Airnews, Inc., U.S. Air Lines, and the Flying Tigers' air lines.

WHEN FOREIGN-made automobiles started coming into the United States market, air freight hauled many of them all over the nation. Bulky cargoes can be shipped economically, Slick has proved.

* * *

LOADING TEXAS blackberries for shipment to eastern markets. Air freight has extended Texas truck farmers' market for highly-perishable products, especially from the Rio Grande Valley. Strawberries and okra growers have found profitable outlets via air freight.

* * *

Revenue Ton Miles Flown	
1950	45,657,828
1949	24,247,012*
1948	25,581,493
1947	21,669,651
1946	11,198,594

*Certificated Sept., 1949. Business up 88% since then.

Air freight eliminates costly crates from many items like these pianos

Relative size of big vans and airplanes used on the air freight runs are shown in this scene from Burbank, Calif., terminal of Slick Airways.

Top Company Officials

Earl F. Slick chairman of the board

Lewis J. Moorman, Jr., board vice-chairman

Thomas L. Grace, president of Slick

W. E. Hollan, executive vice-president

Joseph F. Grant, vice-president - secretary

David R. Stewart, vice-president-treasurer

Slick Airways launched a new concept in aviation history, transporting everything from race horses to okra; rumors persist that it was also used for covert CIA operations. *Betty Slick Moorman Photo Collection*

Vietnam. Aviation buffs still recall that cargo planes were a novel idea dreamed up by two brothers named Slick.

While the airline evolved in its early years, Slick's projects on the ground were booming. In 1947, Slick purchased the Benedum Oil Field from one of his father's competitors for $250,000. The risk paid off, and it became one of the largest post–World War II oil discoveries in the country, later yielding Slick a sales price of two million dollars, which would be equal to more than ten million dollars in 2020, with inflation.

That same year, Slick founded two more institutions that he was sure would move his dream of a "Science City" forward, by conducting diverse research that would improve the world. Southwest Research Institute (SwRI) was established as an applied research facility that would work closely with the Foundation of Applied Research, now in its sixth year of operation. While the foundation would continue to focus on basic research, SwRI was established to secure corporate and government contracts that could produce solutions to industrial challenges in the fields of construction, oil and gas, and chemical and electrical engineering.

Located next door to FAR, the institute shared laboratory facilities and scientists in its early days, and close collaborations would continue over the decades, evidence of Slick's vision for a cluster of scientific institutes working together. The Institute of Inventive Research (IIR) joined the cluster a little later that year, established to encourage creativity by inviting inventors to submit their ideas or products. It was housed in a barn at ESSAR Ranch, and in its first year of operation, Slick received more than five hundred letters and several hundred sample inventions, ranging from hair drying hoods that could be used at home, to a seamless tin can that, while not particularly useful, was quite fascinating.

Slick was very active in managing all three institutions in their early years, making strategic decisions and engaging in long-range planning. By 1948, he recognized that growth of the three organizations had created new leadership demands, and a national search began for a scientist/president.

Dr. Harold Vagtborg was working as the director at the Armour Research Foundation in Chicago, Illinois, when Slick first learned about him through an article he read in *Reader's Digest* when he was stationed in the South Pacific. Vagtborg was quoted as saying that "he could improve anything," and Slick resonated to the message and liked the man immediately. A few years later, he learned that the scientist had become the director of the Midwest Research Institute in Kansas City, Missouri, and Slick pursued him with gusto, hoping that Vagtborg might be attracted by the opportunity to build a science city, rather than maintain one already established research institute.

In the summer of 1947, the two men corresponded about a possible visit to San Antonio, and Slick suggested that Vagtborg bring his family to Texas for the Thanksgiving holiday. He explained that he was in the process of establishing the Southwest Research Institute, promised sunny weather, and said that he would send his plane for transportation.

Slick hired Harold Vagtborg (*left*) to manage his three scientific institutes in 1949. *Courtesy of Texas Biomedical Research Institute*

Vagtborg described the November trip in his book *The Story of Southwest Research Center*. "Tom, on his usual tight schedule, was delayed in Australia, so we arrived in San Antonio twenty-four hours before he did. Mr. and Mrs. Charles Urschel came to our rescue. It was a wonderful experience to meet these gracious people, and it was at this time they told me their ideas of Tom's ambition. Mrs. Urschel expressed to me her great concern that Tom was taking on more than he could possibly handle. She talked with such sincerity and conviction that I told her, 'Mrs. Urschel, I will do everything I can to talk him out of it.' Her response was, 'Don't ever do that. It would be a terrible disappointment to him. We'll just have to work everything out.'

"The next day I met Tom for the first time, and we spent the day at the Cable House, which he had just refurbished, and also had a visit to the laboratory building nearby. It was there in the Cable House that he revealed his plans for the ambition of his life."

Vagtborg was convinced that Slick's vision was so bold in concept that one could only conclude it just might be possible; he agreed to move to Texas. With qualified hands-on management in place, Slick increased his traveling, continuing to funnel a constant flow of ideas for research projects, fund-raising schemes, and potentially important contacts for the scientific institutes from all over the world to Vagtborg.

Between large amounts of time dreaming up scientific projects, considering ideas for new inventions, drilling for oil, running an airline, ranching, and traveling to exotic places, not much time was spent at home. Despite a cozy house in the charming Monte Vista area of San Antonio, and young children, Slick's wife Polly felt isolated and alone. The fairy-tale marriage was experiencing some strain, and Polly began to resent being "molded into the perfect wife." Years later, Nixon family members would sigh ruefully and comment that Nixons never could be molded.

With the same bursts of energy that he poured into every endeavor, Slick made the most of every minute he shared with his children. He planned adventures at the Texas coast and sophisticated outings to see the Rockettes in New York. He scoured catalogs to find amazing toys like gasoline-powered Model-T cars for the boys, and Patty's playhouse with a real doorbell and stove that could bake cookies. He played with his children and their toys with absolute delight. He was the first in the neighborhood to introduce the new concept of outdoor Christmas lights in the yard—creating a life-size Santa Claus, sleigh, and reindeers that drew spectators from all over town.

Although oldest son William (nicknamed Bill by Slick, and called "Wid" by his younger siblings) lived in Dallas with his mother, Slick's letters to his young son were constant and consistent, encouraging him in his studies, urging him to join in summer vacations, and trying to make the difficult environment that divorce creates more tolerable. All four children—Wid, Patty, Tom, and Chuck—established close bonds, remained intensely loyal to their father, and much like the generation before, learned to enjoy the moments of togetherness and to survive the lonelier times.

Slick wrote to his children about his exotic adventures—much the way his father had shared his gypsy life in the oil fields—and he sometimes included Travis Richardson and a few other members of his inner circle in the journeys he made and the discoveries they produced.

"Tom would go off on a trip," Dorothy Richardson explained, "and we'd get a magazine all rolled up with grass seeds. Sometimes bugs would crawl out; and I'd have to burn the whole thing.

"Sometimes the seeds would be pristine; and we made some wonderful advances in grassland innovations in collaboration with the King Ranch."

After a trip to India in 1949, he contacted Robert Kleberg, who was both a family friend and the patriarch of the famous King Ranch, at that time the largest ranch in the world with more than one million fenced acres. He'd seen some plants and grasses on the other side of the world that he thought might be perfect for Texas and other parts of the United States that are hot and dry, and suggested that the King Ranch might test them after the agriculture division of FAR had completed its research.

In addition to their shared interests in cattle and pasture development, Slick and Kleberg both were intrigued by a ranch that was for sale in Cuba in the early 1950s. Owned by an American woman whose husband had recently died, the ranch was reported to be incredibly beautiful.

"Tom wanted to look at a ranch in Cuba—right before Castro took over—and he wanted me to run it if he bought it. I told him I didn't

During his lifetime, Slick owned several private planes that took him on adventures, including this prized Beechcraft Bonanza. *Courtesy of Mind Science Foundation*

want to move there," Richardson recalls. "But I traveled with him to look at it; and I understood his interest. It was beautiful land, with a lovely American widow, in debt, needing rescue. It had all the elements of a 'Tom Slick Adventure.'"

The airplane was low on gas, but refueling in Miami would not be possible. Flights to Cuba in 1950 were not encouraged; the US State Department was aware of the activities of the *comunistas*, and everyone knew that Fidel Castro was mounting troops for a revolution.

"Can we make it to Havana?" Tom Slick asked.

The small group of travelers was quiet as they waited for pilot Jack Barber's response. Slick was flying to Cuba with his ranch foreman and wife, Travis and Dorothy Richardson, to look at a ranch that was for sale on the eastern side of the island. The pilot hesitantly told the passengers that it would be close. Barring headwinds, they would probably make it without refueling. "Great!" Slick replied, "Let's keep going."

Richardson, telling the story in his late seventies, and still a rancher, remembered the landing as "hostile."

"No one was glad to see us," he recalled, "and the airport officials wanted us to turn around and go back home. But Tom just stepped off the plane smiling, made a few phone calls, and pretty soon we were all drinking *Cuba Libres* with the airport soldiers."

Slick's children saw their father work this magic many times. "My Dad had an amazing ability to get things done," his youngest son, Chuck, explained many decades later. "He just did not understand—or listen to—the word 'no.' If anyone said something could not be done, Dad would try it.

"As children, Patty, Tom and I would try to figure out how he could make a few phone calls and have tickets to a sold-out play, or dinner reservations someplace that needed to be booked weeks in advance. And everyone in the 'transaction' always seemed to be smiling. I'm still trying to figure it out!"

Whether it was securing special tickets to Disneyland for his children or landing in Havana when the airport was officially closed, Slick believed that most "no's" could be overcome. Sometimes he used charm; sometimes he used cash; sometimes he used what his friends called a special "magic"; and sometimes he used a combination of all these ingredients to get what he wanted.

In Cuba, he wanted adventure and possibly a new ranch. After stocking up on supplies and hiring two native guides, Slick and the Richardsons traveled in a battered jeep to the far side of the island. The *comunistas* were patrolling the countryside, blowing into conch shells to send messages across the lush Cuban valleys. Richardson was aware of the political drama about to unfold amid the tall coconut orchards and tobacco plantations, far from Havana's elegant architecture and glamorous nightlife.

The rainy season had destroyed most of the dirt roads; the jungle vines and thick foliage had to be hacked back by Cuban workers wielding machetes; it was hot and humid. Every so often, Slick and the Richardsons stopped for refreshment.

"I'd chop down a coconut, and Dorothy would drink most of the milk inside," Richard remembered. "Then Tom would pour rum into the shell and the men would sip our own version of *piña coladas*."

Slick went to Cuba with Travis and Dorothy Richardson; decades later Richardson remembered the homemade drinks they made in the jungle from coconuts and rum. *Travis Richardson Photo Collection*

The ranch was beautiful, owned by a family friend of Slick's who had recently been widowed and wanted to move back to the United States. The spectacular ocean view, the lush grasses that would make wonderful grazing land for cattle, and a "woman in distress" were all factors in Slick's decision to buy it. But Richardson balked at the prospect of moving to Cuba in 1950 to run a new ranch; he urged Slick not to purchase it.

"Tom was disappointed," Richardson said, "especially when the King Ranch decided to make the deal. They took horses and cattle over there and set up quite an operation. A few years later, Cuba took it all. Then I was a hero," Richardson laughed.

Slick was a frequent visitor to the early laboratories at the Foundation for Applied Research. *Courtesy of the Mind Science Foundation*

5

Slick Ideas

1950–1955

He addressed the letter to "The Indian Government, New Delhi, India," added an airmail stamp, and waited for an answer. Tom Slick never considered that the rather vague address, or possible security issues, might interfere with the delivery of his handwritten note:

> *Dear Sirs:*
>
> *My attention has been called to a newspaper article giving some information on a specially devised box to send spawns of fish alive by air to distant places.*
>
> *I am interested in getting some spawn of Monsoor? (spelling?) fish from India if you think it feasible and if they can be transplanted in the United States.*
>
> *Any information you can give me on this matter will be greatly appreciated.*
>
> *Very truly yours,*
> *Tom Slick*

A few weeks later, an official-looking envelope covered in Indian stamps was waiting on his desk. The information he had requested had arrived, and he began to arrange to bring the famous fighting fish of the Ganges to the Guadalupe River in Texas.

A voracious reader, especially of newspapers and magazines, Slick always turned pages with his imagination at full throttle. He searched for anything new or controversial and made quick mental leaps to ways the information might be applied to his scientific research programs, livestock and ranching business, oil ventures, interest in world peace, or some exciting new gift for his young children.

Slick wrote to President Dwight Eisenhower about the need for the United States to remain on the cutting edge of technology internationally; to Aristotle Onassis about tankers; to Norman Cousins, David Rockefeller, and Secretary of State John Foster Dulles about world peace; to construction powerhouse George Brown about building a floating recreation center in New York's East River; and to oilmen Clint Murchison and Sid Richardson about an even more valuable natural resource—water. He knew some of these powerful men personally, but some were total strangers. He never hesitated to contact anybody, with any idea. His wealth certainly contributed to his confidence that those he contacted would respond, but there was more to it than that. He'd learned from his father, Lucky Tom, that "nothing ventured meant nothing gained," whether it was prospecting for oil or making a telephone call.

In 1954, Slick bought a plantation in Mississippi and began to think of crops and livestock that might be introduced there. After a hunting and fishing trip to New Zealand that summer, he wrote to a rancher down under to inquire about the possibility of bringing Romney Marsh sheep to the United States, explaining that "it would be interesting to try them out in this country, and if successful, build up some sort of popularity in the breed." He asked S. C. Sutherland to send him three rams and six ewes, adding that "we would also like to breed up some Cordales or some other type of starting stock. We would want them for meat as well as wool, with meat being perhaps more important than wool in this country. I would appreciate your thoughts."

He became interested in an inventor who claimed to have developed a men's hair tonic that could change the direction that hair grew. His brother-in-law, Jim Nixon, remembered it was an appealing idea in the 1950s, when men coated their hair with thick balms like "Lucky Tiger" to force unruly cowlicks into the smooth, slicked-back look of the times.

Slick found an article about a "Four-in-One Trailer" in the June 1955 issue of *Popular Mechanics* and immediately wrote to the manufacturer in West Virginia for details about its ability to function as a raft, boat trailer, and camper. That same year, he ordered a pair of skunks that had no scent glands as an interesting holiday surprise for his sons, Tom and Chuck. Their unique pets added tremendously to their popularity that year.

Slick put equal thought and energy into gifts for his wife, Polly. In 1954, he wrote directly to Stanley Marcus, the CEO of Neiman Marcus, then the most elegant department store in Texas (which would expand nationally in the 1980s), asking him to help find a perfect gift for her birthday. Ironically, one of his ideas was a set of beautiful luggage, "blue and green plaid," so she would be equipped to travel on some of the adventures he was planning.

Later that year, the couple appeared in *Time* magazine, in a photo with US President Eisenhower. All were smiling and looked happy, but the facts were different. The model of marriage that had worked for his mother and father in the early 1900s was not as successful for Tom Slick nearly fifty years later. While his children accepted his absences and enjoyed the adventures they shared together (as Tom and Betty had done with their father), Polly Slick expected more. In 1955 they divorced amicably, and they remained good friends for the rest of Slick's life. Eventually she remarried, and at times, she and her new husband, Paul Pater, the three children, and Slick would have dinner together. All recognized a complex, important bond and took special care to maintain it.

With the change in his home life, Slick allocated even more time to his research institutes and new technologies in the oil business. In 1955 he met Robert West Jr., a young man who had just earned his doctorate in chemical engineering from the University of Texas. Slick liked West's combination of academic expertise, practical know-how, and creativity, and offered him the job of running Slick Oil. Many decades later, West smiled as he remembered telling Slick that despite his strengths in related fields, he did not have any experience in the oil business. He was hired immediately. Slick used intuitive hiring practices, based more on the applicant as a person than anything else. West went on to supervise innovative new secondary oil recovery processes and other pioneering technologies, and eventually he bought the oil company after Slick's death (renaming it Tesoro) and built it into the first Texas company to be listed on the New York stock exchange, where it was valued at $42 billion.

By the mid-1950s, after more than a decade of existence, Slick's scientific research organizations were beginning to thrive. The first oral contraceptives developed at Southwest Foundation for Research and Education (formerly FAR, until its official name change in 1952) were proving successful, and plans were underway to develop a new primate research colony, recognizing that the close genetics of baboons to humans could yield important medical treatments for the future.

Vagtborg was in place as a full-time director of the Southwest Foundation, the Southwest Research Institute, and the Institute of Inventive Research, and Slick encouraged him to build an expanded management team. In 1955 Vagtborg hired Martin Goland, an aerospace engineer whom he lured to Midwest Research Institute when he was director there. As the new head of the SwRI's technical program, Goland convinced Hungarian Stephen Juhasz, an engineer and editor of *Applied Mechanics Reviews*, to join the migration to Texas as well, and his expertise in fossil fuels research and heat transfer soon elevated SwRI to new levels of engineering. Juhasz, in his nineties in 2001, remembered that the "staff directory filled an entire page, we had two hundred employees." Over the next seven decades, SwRI would expand its workforce to nearly three thousand.

Scientists and inventors who were in the laboratories during the late 1940s and early 1950s remembered many "Slick ideas," and praised the overall spirit of open-mindedness and creativity that was so important to their boss.

"Tom Slick believed that people and ideas were more important than bricks and mortar," seventy-five-year-old Bill Mallow, a materials scientist and inventor at the Southwest Research Institute for more than forty years, said in 2001. "He would stop by my office, and the visit never ended without at least a dozen new twists to what I was working on, and usually some thoughts on what I might do next."

Mallow went on to find the right chemical ingredients for Liquid Paper, a secretary's idea for correcting typing mistakes, and invented a special soil additive to reduce the formation of potholes in aging roads. He also worked briefly on Slick's artificial pecans project.

When Slick realized that a drought in Texas, where most commercial pecans are grown, would affect the cost of the precious nuts, he suggested that scientists at Southwest Research Institute should invent an artificially flavored "pecan," without a shell (since shelling also raised the cost of the natural nuts), that would be an economic alternative for the baking industry.

After about a year of careful mixing, and work on color, consistency, and flavor, the artificial nuts were ready to market. But a drought-free year produced a bumper crop of real pecans, and a new shelling device brought their cost down still more. That year, most of the artificial pecans found their way into holiday baskets at the Institute, along with turkeys, jalapeño jellies, champagne, and other delicacies that Tom Slick gave his employees and friends during the Christmas season. "They would be an odd, but precious, souvenir today," Mallow concluded with a chuckle, "but of course everyone used them to bake cookies during their brief existence."

Slick enjoyed ideas and liked people who had lots of them. Some of his more conservative friends often warned him that he was too open, making him vulnerable to opportunists. He countered with his own views about discovery. "I always believe in taking a positive approach, instead of a negative one where everyone just sits back and says that everything inconclusive is impossible," he wrote in 1952. The frontiers of possibility were where Slick preferred to spend his time, and he never hesitated to give any person or idea his thoughtful attention, however far from traditional thinking or so-called proven science it took him.

"I have found that in many cases highly trained technical men are not willing to even look at claims that are beyond known technical theory," he said, "and I'm convinced that some of the most important advances will come from accidental discoveries."

Through both the Southwest Foundation and the Institute of Inventive Research, Slick received thousands of ideas—some scientific and very exciting, others totally unscientific but also very exciting. He loved the profound science behind an idea that arrived in 1954 from Dr. Harlow Shapley at the Harvard College Observatory, suggesting a design for an electronic telescope, which would convert light to electrons, multiplying and converting back to light. A similar concept found practical medical applications about the same time, when Westinghouse developed and marketed an x-ray image intensifier that enabled fluoroscopes to produce bright, visible light output for diagnostics.

That same summer, Slick wrote to the Woods Hole Oceanographic Institution to learn more about their new device to measure the speed and direction of ocean currents while a ship was moving. "I was wondering if the same instrument could give you a true ground speed indicator in an airplane, for which it seems to me there is a very great need," he told the research group, offering to explore a cooperative development between the Woods Hole institute and the Institute of Inventive Research.

The 1950s produced extraordinary ideas and inventions that are such an accepted part of today's twenty-first-century life that it is difficult to imagine how remarkable they were at the time. In 1953, AT&T began to lay the first telephone cable under the Atlantic Ocean, and new refrigeration technologies made it possible to develop a home freezer. Long-distance telephone service and frozen foods were novel ideas less than hundred years ago, and Slick kept a notebook as exciting new discoveries were made. He tracked as many marvelous inventions as he could, suggesting linkages to the four research institutes he had founded over the past decade. All part of Slick's overall vision of a "Science City," they were like siblings, each with its role in the family. First born, Southwest Foundation was chartered to conduct basic research. Its early work produced new discoveries in agriculture and livestock breeding, including new drought-resistant grasses and artificial insemination, and later, breakthroughs in medicine, including oral contraceptives, new treatments for heart disease, important advances in genetics, and eventually vaccines for Hepatitis A, B, and C and HIV/AIDS, and cutting-edge research of infectious diseases like Ebola and Hanta.

Perhaps because Slick was such a young man—just twenty-five years old—when he "fathered" this first scientific foundation, he remained especially proud of it throughout his life. A few scientists from its earliest days remembered how different scientific research was at ESSAR in the 1950s.

Dr. Pemmaraju Rao was one of the first scientists at FAR; his career spanned more than fifty years and produced major discoveries in organic chemistry and hormone research. *Courtesy of the Mind Science Foundation*

"After completing a Fulbright Scholarship at Rochester Medical Center in New York in 1954, I returned to my home in southern India," Dr. Pemmaraju N. Rao explained in 2001, gazing toward the multimillion-dollar biocontainment laboratory visible just outside his office window at the Southwest Foundation for Biomedical Research, contemplating the changes he has seen in the surroundings during more than four decades. Dr. Leonard Axelrod, a colleague at Rochester, had written to Rao, urging him to come back to the United States to work with him at a new research facility in Texas.

"When I arrived in 1956 from India, I discovered a ranch house in the middle of nowhere, and I was totally shocked," the veteran scientist laughed. But complete scientific freedom and the confidence that emanated from Tom Slick, along with the promise of more modern facilities soon, were enough to keep the foundation's first small staff of six scientists passionately at work in their somewhat "wild west" setting.

Rao and Axelrod made important discoveries in steroid research, which eventually led to new understandings about hormone metabolism, its link to cancer, and some of the antihormones used today to treat breast cancer and other tumor inhibitors.

Rao remembered finding a snake in his laboratory refrigerator, and Joseph Goldzieher, one of the world's pioneers in contraception and reproduction research, recalled that his un-air-conditioned work space was so hot that his team had to work at night to avoid being anesthetized by the ether they were using in their experiments. Both men agreed that Tom Slick's offer of total freedom to design and conduct their research was more appealing than the state-of-the-art facilities they might have had elsewhere.

Goldzieher was a 1940 graduate of Harvard University and received his medical degree from New York University College of Medicine. He balanced a private medical practice, teaching, and research, for more than sixty years and received numerous awards for his discoveries, most notably the development of oral and injectable hormonal contraceptives.

Slick was convinced that population control could offer the world an important opportunity for breaking the cycles of poverty and conflict. Goldzieher remembered their conversations about the positive impact of contraception and family planning long before the topic gained any public awareness. During trips to Africa, India, and other developing nations in the late 1950s, Slick always mentioned the work underway at the Southwest Foundation with pride and an optimistic confidence that it would make a positive difference in world economies. Decades later Goldzieher saw his discoveries in endocrinology and hormone research honored by organizations around the globe, including the Women's Health Initiative; today, oral contraception methods are practiced around the world.

"Tom Slick was interested in me because I was from the other side of the world," Rao said with a smile. "He was fascinated by the mystical aspects of life, and by 1958, he had visited India, Tibet, Nepal, and Bhutan. We had great conversations about paranormal phenomena, and he was organizing a new scientific institute to study them."

The Mind Science Foundation was established in late 1958, and in 2004, Dr. Pemmaraju Rao became one of its most respected board members. While still an active scientist at the Southwest Foundation for Bio-

(*Left–right*): Bill Rhame, Betty Moorman, and Slick came up with the clever idea of establishing an exclusive private dining club where all proceeds would support scientific research. *Betty Slick Moorman Photo Collection*

medical Research, Rao, like Slick, dedicated his life to scientific discovery, without ignoring those mysteries that may not always have a scientific explanation.

About the same time that Rao, Axelrod, and Goldzieher arrived in Texas to work at the foundation, Slick met a Harvard Business School graduate named Bill Rhame through mutual friends in New York. Rhame had graduated in record time, with honors; Slick was impressed by his intelligence, clever wit, and readiness to take risks in the business world. Deciding he would be a perfect manager for his multifaceted business interests, Slick lured him to San Antonio as well, organizing Texstar Corporation as a holding company for all his endeavors, and making Rhame its new president.

Later that year, Slick's sister, Betty Moorman, suggested an innovative way to raise money for her brother's favorite research facility, and Rhame assisted with the rather tricky financial planning. Jesse Oppenheimer, a family friend and respected attorney, added his expertise to the evolving idea that eventually would provide more than $1 million in annual contributions to the Southwest Foundation.

The Argyle Club opened its doors in 1956, furnished with Betty Moorman's and her friends' extra chairs and tables; it became a major source of funding for Slick's research foundation. *Courtesy of Texas Biomedical Research Institute*

Moorman gathered her best friends and asked them to help her design a unique and exclusive dining club, where all dues and profits would go toward supporting research at the foundation. The city's most elegant hostesses joined forces with the sharpest legal and business minds, and the Argyle Club was born in 1956.

Located on a quiet street in one of San Antonio's most beautiful residential areas, the club had once been a boarding house and a bakery. Its southern charm includes a magnificent lawn and garden, an old-fashioned veranda framed by high colonial pillars, several elegant and intimate dining areas, and a long waiting list for membership.

Oppenheimer's creative legal thinking produced a plan that would enable a club to be established as a nonprofit entity to support the foundation, built with tax-deductible funds. The old boarding house was purchased for $32,203, a renovation plan was designed, and a membership strategy that was advantageous to its members as well as the foundation was created. Since the plan was so novel, a hearing to consider its validity was held at the IRS offices in Washington, DC, and Oppenheimer was successful in his negotiations to establish the Argyle as a 501(c)4 nonprofit entity.

Oppenheimer was on the club's first board of directors, along with Moorman, Rhame, and other civic leaders including socialites Peggy Becker and Elizabeth Urschel, banker Tom Frost, and arts patron Margaret Tobin. Oppenheimer's recollections of early meetings with some of the city's first members included decisions to furnish the club with "extra things" stored in their own attics, and he recalled that the first Christmas party at the Argyle in 1957 was a huge success.

"I learned an important lesson for the future," Oppenheimer remembered with a smile. "As we went around the Board table, voting on whether or not to have a party, each person said 'oh, no . . . there's too much going on at Christmas and the last thing we need is another party to go to.'

"When we got around to Peggy Becker, she rolled her eyes and in her very appealing way said, 'Charlie and I don't have too much to do at Christmas time. We have said "no" so many times that we don't get to say no anymore!'

"Her statement has guided me in many situations," Oppenheimer said, adding that he never heard Tom Slick say "no" either.

Today, other nonprofits around the country have followed the innovative fund-raising model, establishing exclusive social clubs that direct all profits to their philanthropic missions.

Slick was a tireless fund-raiser. He wrote to his family, friends, and complete strangers, urging them to support his first research venture. Like a proud father, he listed its accomplishments and shared his dreams for its maturity. In a letter to his mother, Berenice Urschel, Slick explained the new changes in tax law as positive for charitable giving and suggested a "matching grant" idea, urging her to match her brother-in-law, Arthur Seeligson's year-end contribution in 1957:

> *Old Arthur, who as you know is plenty smart about taxes, and throws money away just like glue, has offered to give the Foundation $12,000. This is about as strong a recommendation that it makes money for him as I can think of.*

Still another innovative fund-raising strategy emerged in the form of memorial gifts. Today, contributions made to a nonprofit organization to honor a family member or friend who has died are very common; in the 1950s, they were controversial, viewed by some as inappropriate. Flower merchants were especially critical of this new "competition," but the Southwest Foundation promoted the memorials as more meaningful.

In a letter defending the foundation's position, Slick admitted that "my personal feeling is that every expense connected with funerals is a great waste which should be put to more constructive purposes. . . . I feel very strongly about it, as I have seen some cases where people who very badly needed the money were talked into spending it on useless funeral expenses at a time when their sales resistance was low. For me, flowers are something that should be used for happiness and happy occasions. For a memorial, I personally cannot help preferring something more constructive."

Despite his strong defense of the new program, he was sensitive to how others might feel and was careful to say often that preferences always are personal and should be honored as such.

His attention to personal preferences also extended to his hiring practices. He believed that finding the right people for any job was the real key to future success, and he used a special "technique" to lure the very best to his team. He called it "the red car offer."

To attract the scientists, oilmen, ranchers, businessmen, and others who joined Slick's different endeavors, he first gave serious thought to each candidate's particular personality and desires, understanding that every human being is unique, with specific needs and hopes. When Slick discovered that mystery in each potential employee, he offered the very thing that made the job irresistible. Sometimes the lure might really be a red car, but usually it was something less tangible and more meaningful.

In ranch foreman Travis Richardson's case, it was the opportunity to realize his longtime dream of running a cattle operation. In chemical engineer Bob West's case, it was the challenge of hands-on experience in the oil business and innovative experimenting with new recovery methods. In Dr. Harold Vagtborg's case, it was the chance to build a bold new research endeavor rather than maintain a more established laboratory.

In Bill Rhame's case, it was the opportunity to combine business with extremely creative risk-taking.

Slick had an unusual sensitivity to dreams and longings, both his own and those of others. When he found human talent and spirit that he liked, he often reshaped the job description to add that person to his team.

It did not always work. Both of his wives possessed the talent and spirit he liked, and they were intelligent and beautiful as well. But neither could adapt to the "reshaped job description" that was part of marriage to Tom Slick. While he tried to satisfy their dreams of a conventional home life, his own dreams bubbled up from a deep well of curiosity about the world, and exotic people, out-of-the box ideas, and mystic landscapes danced in Tom Slick's mind. New rumors of a diamond pipeline in the Amazon jungle, and separate reports of a shaman's extraordinary abilities were about to lure the mystery hunter away from home again, to faraway places.

Stories of a "diamond pipeline" and shamanic healers in British Guiana lured Slick to the jungles in 1956. *Catherine Nixon Cook Photo Collection*

6

Diamonds and Death

1954–1956

In the spring of 1954, Tom Slick met an adventurer named Leonard Clark in San Francisco. The two discussed their shared interest in exploration, and Clark gave Slick a manuscript describing conditions in the Amazon region of South America, where Slick thought there was potential for oil exploration. Soon after the meeting, Clark wrote to Slick, inviting him to participate in an expedition to British Guiana (now Guyana) in the Amazon to investigate oil indications, mineral deposits, and Indian medicines used by native *curanderos.*

Although not able to accompany Clark, Slick was intrigued and wrote back about the Indian medicines with enthusiasm.

> *If you find any medicines that seem to be good, be sure to bring out sizable samples so that we can have our scientists confirm their effectiveness. Also, if you find any oil seeps, bring out some bottle samples together with careful descriptions and locations.*
>
> *Letter from Tom Slick to Leonard Clark*
> *April 26, 1954*

Slick explained that while Clark would be on expedition in South America, he would be in India studying "the superior control of the mind which they claim the Indians have achieved," followed by a journey to the Himalayas to investigate reports of the "abominable snow men."

The following year, Clark sent Slick a copy of *Gold, Diamonds, and Orchids*, by William La Varre. Clark knew La Varre slightly and was interested in his claim to have discovered a "diamond pipeline" in South America. Slick read the book and sent it on to Bob Kleberg, a close friend, and the owner of the famous King Ranch, then the largest ranch in the world. For once, Slick was a bit skeptical, noting that "this fellow certainly writes an interesting adventure story which he tells me is true . . . it is a little hard to believe in some of the details."

Leonard Clark continued to correspond with Slick, sending him reports of intriguing discoveries, and with time, Slick's curiosity overcame his skepticism.

In 1956, he organized his British Guiana Expedition with a budget of $13,000. The diamond-hunting team included Slick, who would contribute $10,500 of the funds needed; Ted Macklin, an associate of Leonard Clark's from California, who would contribute $2,500; Leonard Clark, who would serve as expedition field leader and be paid $500 a month for two months; Macklin's son Dick, who would serve as deputy field leader and be paid similarly; Dr. Mather, a geologist on leave from the Southwest Research Institute, who would be paid his regular salary; and Jack Barber, the pilot of the private plane that would transport the team and its equipment and supplies to South America.

As the departure, planned for early November 1956, approached, Slick decided to add another team member—his ranch foreman, Travis Richardson. Explaining that he was "a tough young fellow, a good shot, and someone I trust," Slick knew he would add common sense to the expedition, and that he could investigate South American ranching opportunities during the same trip.

Slick would fly commercially to Buenos Aires and Rio de Janeiro on private business, then on to Caracas to join the rest of the expedition team, which was scheduled to fly straight from Texas to Venezuela.

"Our plane from San Antonio to South America had guns and ammo and equipment, but no permit," Travis Richardson recalled. "We made it to Maracaibo, where we were grounded, and the plane was sealed. I begged them to hold the plane 'in transit' while I flew to Caracas to meet Tom and tell him what had happened.

"Once I found Tom in Caracas, he made a few phone calls, and everything was fine. The grounded plane in Maracaibo got clearance and flew

In search of treasure and magical healing potions, Slick traveled up the Potaro River with Richardson and explorers Leonard Clark and Ted Macklin. *Betty Slick Moorman Photo Collection*

on to Georgetown, British Guiana. Tom and I met the rest of the group there, and we got ready to go into the jungle."

Georgetown was a wild place in 1956, much like the gold rush towns of California and Colorado a century before. Richardson remembers that it was filled with adventurers looking for diamonds, gold, or any other business that might mean money.

"I remember one young man from New York City was in Georgetown, said he was in the diamond business, asked me who our expedition's 'money man' was. That was an interesting thing about Tom—he always blended in with everyone else; no one ever knew that he was a millionaire."

The team went deep into the jungle, and Slick was as excited as a small boy at Christmastime as he collected samples of natural herbal medicines and cures from the Indians. While Slick enthusiastically courted the shamans, urging them to share their secrets, others on the expedition were not as trusting. The men took turns keeping a watch, sleeping every other night, not at all certain what these foreign men dressed in loincloths might do. "It was like the movies," Richardson says. "The men had painted their faces and bodies and were very vain; the women were plain-looking and seemed to do all the work."

The search for the "diamond pipeline" was not as successful as the investigation of traditional medicine, but Slick's enthusiasm was not deterred. He was sure that Clark and Macklin would find the mystical place described in La Varre's book, *Gold, Diamonds, and Orchids*, where the natural rock and soil conditions created an unnatural "pipe" for the formation of the precious stones. They planned to stay on in British Guiana, exploring the jungle, while Slick, Richardson, and the geologist Mather would return to the United States.

"When we came back from our first jungle search for the pipeline, the diamond salesman from New York was still in Georgetown, talking about a gold deal. Tom had to get back to Texas, so he didn't bite, otherwise he might have. That was always a danger . . . he was curious about everything."

Getting back to Texas did not prove simple. The plane would have to make two flights from Georgetown to Caracas since not all the team members and equipment could fly in the plane at one time. As the first flight took off from the grass strip, the plane's wings dipped dangerously, and it was obvious to everyone on the ground that something was terribly wrong. An experienced pilot, Jack Barber realized that he had engine trouble and tried to bring the plane back down, crashing into one of the giant anthills that stood several feet tall on the turf just beyond the airstrip. No one was hurt, but the plane was badly damaged.

Richardson, scheduled to go on the second flight, was on the ground and watched in horror as the plane with Tom Slick aboard crashed to the ground. Later, he would ask his boss if he'd been frightened. "The answer was 'no,'" Richardson says, "Tom explained that he believed in predestination, which really makes fear nonexistent."

According to some friends who remembered the South American airplane crash, the US State Department had warned Slick that there was a lot of Russian activity in British Guiana; some believe that sabotage may have been the cause of the crash on the grass airstrip in Georgetown. The engine only had 291 hours on it and the plane had been completely serviced before it left Brownsville, Texas.

By the end of 1956, Slick and Richardson were back in Texas, but Macklin and Clark continued to hunt for the diamond pipeline, and now for gold as well, renaming the expedition the Polvo de Oro (Gold Dust). They kept Slick apprised of their adventures, but by early 1957, the "money man" was ready to close down the operation. He wrote to his explorers, praising them for their hard work and expressing some disappointment in William La Varre's character and published accounts of the pipeline.

> *Sounds like you've had some real adventures. I was proud to hear that you were able to follow the deductions of Ted (Macklin), but of course it was disappointing that you found nothing at the final site. I don't know if Lavarre is just a liar or what. We have been trying to contact him since my return but so far have not been able to get in touch with him.*
>
> *. . . we had better cross off that part of the project.*
>
> *January 7, 1957 letter to Dick Macklin and Leonard Clark*

The letter continued for five pages, outlining all sorts of other potential deals that Slick was developing in South America. He offered to pay Macklin and Clark to be his "scouts" and instructed them to further investigate the gold property deal (Polvo de Oro) as well as to follow up on the salvage efforts for the crashed airplane and to check into partnership potential in a proposed air service trading company that was being developed in British Guiana.

Macklin and Clark stayed another year carrying out these instructions and exploring the jungles. Then tragedy struck as they prepared to return to the United States at the end of 1957. As a last "holiday" adventure in the jungle they had come to know well, the two young men decided to take a canoe trip down the ferocious Paraguay River. After a plane crash, jungle encounters with tribesmen who still practiced the art of shrinking the heads of their dead enemies, and a life-threatening snakebite, they felt their river trip would be an easy jaunt. The water volume was especially high that year, due to torrential rains, and their inexperience did not prepare them for the strong current that swept their canoe over the falls.

The deaths of Leonard Clark and Dick Macklin brought the South American operations to an end, and Tom Slick had the difficult job of contacting the young men's families and expressing the sadness and sympathy that he felt. His own experiences of loss, especially his father's death at a young age and its effects on his entire family, made Slick especially sensitive to grief. He was quick to call and offer sympathy; he followed up with thoughtful notes for many months; he put aside his disappointment that the Macklin/Clark manuscript from the South American endeavor was never recovered. As was his way, he turned his thoughts to the future, to ways he could introduce the Indian medicines to his scientific labs for study and to a way he could use the few diamonds found in the Amazon to create something meaningful.

After two years of searching in the jungle of British Guiana, five small stones had been retrieved; one was an unusual greenish color. Slick asked Bob Haack, a jeweler in Wisconsin, to make a ring for his ten-year-old daughter, Patty, using the diamonds. His letter asked the jeweler to use the green diamond in the center and instructed him to make the ring "as simple but as effective as possible . . . something she could probably wear later on too." He added a postscript that Patty's birthday was approaching and asked that it be sent in time to surprise her. She never knew the underlying cost of those diamonds, but she treasured the ring, presented by a father who moved easily and often between adventures and always came home bearing gifts for the people he loved and the research efforts he championed.

From the Amazon, Tom Slick brought back jewels for his little girl, dart guns and a shrunken head for his sons, and a pet monkey for his brother, Earl. He also brought bundles of plants, vials of strange liquids, and stories collected from shamanic healers back to the Southwest Foundation where scientists would try to differentiate between magic and science.

Slick had seen firsthand the miraculous effect one medicinal plant had on a snakebite. The Indians had used delicate poison darts to hunt as effectively as any big game hunter, and the tiny, exquisite shrunken heads he brought back to Texas certainly were proof that the jungle shamans knew things that Western doctors did not.

After taking his specimens to the laboratory at the Southwest Foundation, where scientists could study their chemical compositions, Slick invited his ex-wife's father, Dr. James W. Nixon, to come see them. Nixon had earned his medical degree from the University of Pennsylvania in 1919, studied orthopedic medicine in Vienna, Austria, in 1920, and eventually practiced surgery in San Antonio for fifty years, retiring at the age of eighty. Unlike many Western physicians, he did not scoff at

Slick brought back samples of medicinal plants from the South American jungle for analysis at the Southwest Foundation for Research and Education. *Courtesy of Texas Biomedical Research Institute*

traditional medicine; in fact, he had grown up with it and respected what it might teach the modern world.

"The difference between a poison, a medicine, and a narcotic is only one of dosage," he said as he examined the samples Tom Slick's expedition team had collected in the rain forests of Guyana. As a young boy, Nixon accompanied his father—a combination of country doctor and Texas rancher—on medical calls in a horse-drawn buggy. He never forgot that he'd seen homemade potions work as well as pills, or his father's willingness to listen to the *curanderos* who had learned "magical healing" in Mexico. He told Slick about an old mestizo who had arrived at the family ranch with a burro loaded with brush. The Indian claimed that the plants could cure diarrhea, which was a chronic problem in Texas and Mexico in the late 1800s. Nixon was a scientist at heart, and like the majority of the population in those days, also suffered from dysentery. He decided to experiment. Following the old man's advice, he brewed a tea from the *chaparro amargoso* ("bitter bush") and tested it on himself. Impressed by his miraculous cure, he sent the plants to London's finest pharmaceutical house, Sharp & Dohme, and in 1898, wrote the first treatise on a cure for amoebic dysentery. The curative compounds from bitter bush made drugs that are still used today.

As Nixon related this childhood memory to his ex-son-in-law, Slick felt his excitement mounting. While in British Guiana, he actually had tasted a variety of bitter bush. Late one afternoon, after the most remarkable fishing trip of his life, where large fish were caught by dainty poisoned darts, blown from the mouths of Indian fishermen maneuvering their canoes over rocks and rapids, Slick listened as the old shaman who had accompanied the expedition pointed to a thin bush, with gray bark and small red flowers, explaining that it was used to treat fevers and diarrhea. It was so bitter that Slick had followed it with a rum chaser, which he later learned was a century-old custom.

The curative powers of South American bitter bush were first documented during the colonial period, when a famous black medicine man escaped from a slave colony in British Guiana and managed to reach the Netherlands. He brought plants and herbs from the jungle with him, and his power to cure diarrhea caused by intestinal parasites soon was legendary. His most potent concoction was a tea made from "very bitter wood."

About the same time that the country doctor from Texas was writing his treatise and introducing a cure for diarrhea in the southwestern United States, an escaped slave was introducing a South American version of "bitter bush" to Europe.

In British Guiana, patients soon learned that the very best delivery of the medicine occurred when a cup was carved from the bitter wood and filled with rum. The liquor absorbed the curative compounds from the wood, and the drink tasted far better than the bitter tea.

Slick was intrigued by ancient medicinal potions and healing rituals for several reasons, the simplest being his belief that modern medicine could learn from them. But his growing fascination with the untapped power of the human mind also led him to explore them as possible gateways to altered states of consciousness.

Natural hallucinogens were a part of many cultures in the remote places that Slick traveled. From the jungles of the Amazon to the Kalahari Desert, from the volcanic plateaus of Mexico and Central America to the high peaks of the Himalayas, Slick had interviewed healers about their "out of body" experiences, of transformations into jaguars, lions, flying tigers, and other creatures, in order to travel more quickly through mysterious realms.

"What is it like?" Slick wanted to know.

"You feel like the wind," a *curandero* in Veracruz told him. "I feel my lion-hair growing," a Jul'hoansi Bushman in Botswana explained. "It is like dying," a South American shaman said.

Facing one's own death comes up often as part of the rigorous training required to become a healer. And unlike Western civilization's avoidance of the subject, the mystic worlds embrace death as a part of life. Slick loved to discuss the subject with friends, often asking what they thought happened after death, always alert for new ideas on the subject.

Ethnobotanist and anthropologist Wade Davis, author of *The Serpent and the Rainbow*, *One River: Explorations and Discoveries in the Amazon Rain Forest*, and *Light at the Edge of the World*, has spent more than forty years researching and writing about these topics, first as a student at Harvard, and as an Explorer-in-Residence at the National Geographic Society. According to Davis, some of these mysteries can be dangerously potent, despite their tempting promise to take the curious "on a thousand flights of spirit."

Flights of spirit appealed to Slick, and he always wondered what was on the other side. He didn't hesitate to taste the bitter concoctions he was offered in the jungle, but according to his diary, none transported him beyond his hammock.

Noting that the Amazon possesses more varieties of plants and animals than any other place on earth, Slick looked for the *curura*, a giant toad that weighs as much as seven pounds, and the incredible capybara, a rodent that weighs 120 pounds. But despite these exciting possibilities, hidden in the immense green canopy of jungle, another more elusive creature and reports of more mystical healing phenomena on the other side of the world had won Slick's ever-searching attention.

When he left his young adventurers, Dick Macklin and Leonard Clark, in the Amazon to collect as many medicinal specimens and diamonds as possible, he already had shifted his focus to the Himalayas, where mysticism, magical healing, and the Abominable Snowman awaited him.

Science Hunts 'Snowman'

By TOM SLICK
(Copyright 1958)

I believe in the existence of the Abominable Snowman—the so-called "legend" of the high slopes of the Himalayas.

Recently I told this to a doubting friend. When he scoffed at my belief, I told him that I thought the Yeti—as the Sherpas call the Snowman—would be found before the end of 1958. And that the creature would prove to be higher up the evolutionary scale than the ape. And, further, that if I were proved wrong I would donate a thousand dollars to his favorite charity.

Before any mistaken conclusions are drawn, let me emphasize that this does not signify that I take the matter lightly. Far from it. Indeed, it indicates how nearly positive I am in my own mind that the Yeti exists as a humanoid creature. The search for it is surely a scientific project of major importance which could add immeasurably to our knowledge of mankind.

LEADER—Tom Slick of San Antonio is official leader and co-sponsor (with Kirk Johnson of Fort Worth) of search for the Yeti.

Indeed, I have myself made journeys to the Himalayan area. Last year, I sponsored and led an expedition there in search of the Yeti. And, at this moment, the Slick-Johnson Nepal Snowman Expedition is trekking into the largely unexplored, uninhabited Upper Arun River region of the high Himalayas on a new quest, using fresh and exciting methods, equipment and ideas that sprang largely from last year's unsuccessful attempt to present a Yeti to a still-doubting world.

I believe that today, next week, or several weeks from now a report may come out of that ancient wilderness telling that an expedition member has at last come face to face with a fierce and hairy ape-man, eight feet tall.

My interest in the Yeti was first aroused a few years ago when, on a visit to India, I heard vague reports of an ape-man creature roaming the snowy slopes of Nepal, Tibet and the surrounding highlands. In 1954, when a British expedition spent months in a vain, but in some ways encouraging, search for the creature my interest deepened.

Then, in 1956, I made a reconnaissance of my own into the fringes of the high Himalayas. I talked to caravan people coming out of the mountains and to other natives who claimed to have glimpsed the Yeti. Still doubting, I asked these people to describe the creatures.

AT ABOUT this same time Peter Byrne, an Irish journalist and hunter-explorer, was making a reconnaissance of his own in the Sikkim area of Nepal. We met and exchanged accounts of our investigations. Byrne, who speaks Nepalese, had kept verbatim records of his talks with natives. From his notebook, he read me the following question and answer sequence:

Q. "What do you think it is—a man or an animal?"

A. "It is a hairy man. It walks on two feet as a man does."

Q. "What does it eat?"

A. "The same as the yaks (wild oxen)—trees, grass, bushes."

Q. "What does it look like?"

A. "Like a big man, with hair all over. The hair is red, but we believe it changes color in winter. It has a pointed head."

"Q. Does it kill animals or man?"

A. "Yes. A man was killed in a village near Punakha last monsoon time. The Yeti broke his neck and tore his clothes off. His relatives found him, with tracks of the Yeti in the earth all around the body."

These answers tallied closely to those I had gathered in my own interrogations. Further, the men I questioned said the creatures were not bears or apes, which they would recognize, but walked erect and had no tails. These descriptions contradicted the theories of some western scientists who have been trying for years to explain the Yeti "legend" in terms of mistaken identity. The creature, say these conservative scientists, is probably a snow leopard. Or a langur monkey. Or a Himalayan red bear.

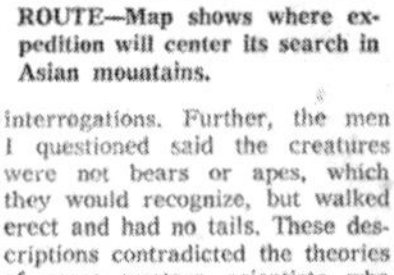

ROUTE—Map shows where expedition will center its search in Asian mountains.

HE'S A BIG BOY—Artist's conception of the giant Yeti dwarfs man. Descriptions put him at eight feet tall. Some reports say Yeti has killed Sherpass without provocation.

HOWEVER, in recent years strange footprints have been found and photographed just above the snowline of the Himalayas—footprints up to thirteen inches long and nearly half that wide, with no indentation of a human arch but more closely resembling the track of man than any animal.

Neither the tracks of the snow leopard, the langur monkey, nor the red bear fit that description. And the bare footprints of man himself are different in shape, shorter and less chunky.

By the time I was ready to return home to San Antonio I had become convinced that the existence of the Yeti as a missing link between ape and man was a scientific long shot worth real investigation. Besides, the mystery challenged my love of adventure. At its lowest level, this would be a chance to hunt a rare and probably dangerous animal — perhaps the most dangerous alive.

I made up my mind to finance and lead an expedition for the following year on a quest for the Yeti. Peter Byrne agreed to join me.

In that we did not actually see, photograph or capture a Yeti, the results of that expedition were negative. But we did find and photograph Yeti footprints. We did follow three sets of the creature's tracks until we lost them on rocky terrain or had to give up due to failing light. And we did collect 15 eyewitness accounts of sightings.

Further than all that, by showing photographs of apes, monkeys, bears and an artist's drawing of a prehistoric ape-man to these eyewitnesses, we asked them to choose the ones that most nearly resembled the Yeti. The unanimity of choice was impressive. Details of these findings will be related in a later article.

Last year, also, the first Russian expedition to the Pamir Mountains, on the far side of the Himalayas, reported sighting a Yeti. Soviet scientists A.C. Promin, of the University of Leningrad, said he watched the creature for five minutes and that it was "stocky, having long arms and a body covered with grayish-brown hair."

BUT APPARENTLY the Yeti is as controversial a creature in Russia as it is in the West. In February, a group of Promin's colleagues wrote a letter to Izvestia declaring they did not believe his claim of a sighting.

My own belief, however, has strengthened with each new scrap of evidence. Now I am almost sure that the creature does exist as an as-yet undiscovered anthropoid type. All that remains to make that belief 100 per cent is the photographed proof of a sighting or the actual capture of a Yeti. If such proof comes this year the whole world will be the richer for it.

From childhood Slick had been fascinated by undiscovered creatures; he wrote an editorial for the *San Antonio Express* newspaper in 1956 before his first Yeti expedition. *Courtesy of the Mind Science Foundation*

7

The Search for the Yeti

1955–1960

"Only a *pukka shikari* (true hunter) will ever catch a Yeti," Tenzing Norgay, the world-famous Sherpa who accompanied Sir Edmund Hillary to the top of Mount Everest, told a group of near-frozen trekkers gathered around a blazing fire at Rathong Camp in Sikkim in the winter of 1956. He explained that the creature was always on the move and very cunning.

According to Tenzing, his father, a yak herder in Tibet, had seen a Yeti years ago in the mountains. It had climbed on top of his hut one night and had stayed there until driven off by the old man's yak dung fire. Tenzing also told the group about a wealthy Texan named Tom Slick who wanted to mount a full-scale expedition to track and capture a Yeti.

A twenty-four-year-old Irishman named Peter Byrne felt goosebumps of excitement as Tenzing Norgay told his Yeti stories and mentioned a man who not only shared the dream of discovering the elusive creature but also had the financial means to do it. Leaving the camp the very next day, Byrne trekked down to Sikkim, retrieved Tom Slick's address from Tenzing's wife, and immediately wrote to offer the oilman his services. He stressed the fact that he already had the necessary permits from the Nepalese government and that he spoke the language. Slick's reply did not arrive for several months because he had spent the winter and early spring in India, just a few hundred miles away from an anxious Peter Byrne, who was ready to switch from guiding tiger hunts for wealthy travelers to searching for the Yeti in his favorite part of the world, the high Himalayas.

> *June 22, 1956*
>
> *Dear Peter,*
>
> *I received your letter and was very interested in it. I am intrigued with the idea of making another expedition—a really active one this time—to try to capture a specimen of the Yeti. I agree with you that your permission to go into that area is an important element; also your knowledge of the language would be very helpful. In regard to the permission, my thought is that we should make it a joint expedition with the Sherpas and the Nepalese so that they would get part of the credit if anything was accomplished.*

Tom Slick shared his research on the Yeti and thoughts about where it might be most likely to live, suggested other possible members for the expedition, and outlined his plans for financing the adventure.

He reminded the young adventurer that three major discoveries of the twentieth century had convinced some biologists and zoologists that there were "fantastic creatures" still to be found. In 1903, the mountain gorilla was seen for the first time in the mountain ranges of Central Africa; the okapi, a short-necked African animal that looks something like a donkey-zebra hybrid, was found soon after; and in 1938, just as Slick was graduating from Yale with honors, South African fishermen caught a coelacanth in their nets. This last discovery had been called the most significant zoological find of the twentieth century since its origins predated the dinosaurs and it was thought to have vanished into extinction nearly sixty million years ago.

Slick had read articles by popular science writers like Willy Ley, William Winwood Reade, and Philip Henry Gosse, whose stories about lake monsters, giant squid, and huge sea serpents fueled his curiosity. When Belgian biologist Bernard Heuvelmans named a new area of research "cryptozoology" and described it as "the science of hidden animals," Slick was intrigued. He agreed with Heuvelmans that "hidden" was preferable to "unknown" since such animals are typically known to local populations, well-described in their storytelling traditions and oral histories.

The stories of "snowmen" in the Himalayas had fascinated him as a boy, and more current reports had excited him so much that he was planning to mount an expedition. When Byrne contacted him, Slick was ready.

He asked Byrne to contact Cathy McLean, a girlfriend who was living in India, explaining that she knew the expedition plans and Slick's unique personality and would be able to assess whether Byrne would fit in with the expedition team.

Cathy had known Slick for about a year; they had met the spring before in Palm Beach, at a house party revolving around golf and girls. Her beauty and intellect attracted the recently divorced Slick, but her surprise announcement that she was moving to India the following week absolutely captivated him. Soon after the oilman returned to Texas, letters began to arrive from Cathy—from Bombay, New Delhi, and Calcutta—describing the people, parties, and mysterious phenomena that she encountered.

By the fall, she had met the Maharaja of Baroda, through Piloo Modi, a friend and member of the Indian government ministry. Her petite build and pale, almost translucent skin were completely foreign and very alluring; the Maharaja brought Cathy to his palace like a delicate treasure, a porcelain doll he had discovered, a special jewel. During all the years that Cathy lived in India, she always was welcome in Baroda; it was her second home. And what a home it was—an architectural treasure with intricately carved walls and delicately mirrored ceilings, surrounded by lush gardens that smelled of gardenias and other more exotic, unknown scents, and fountains that bubbled with memories of India's majestic past.

When Tom Slick visited India in the spring of 1956, Cathy took him to Baroda, where the Maharaja and his wife welcomed them, offering them the European luxuries of electricity, telephones, and champagne, coexisting with ancient traditions and opulence that Indian royalty had celebrated for centuries. The results were unforgettable—elephant rides, a tiger hunt under a full moon, stories of mysterious *yeh-tehs* in the mountains to the north, and amazing demonstrations by swamis who could walk on hot coals without burning their feet.

Who could resist the magic? Certainly not Tom Slick, whose journeys to India's frontiers of mystery had just begun. His "Snowman hunt" planned for the next winter would be just the beginning of a five-year adventure to unknown realms—primitive Nepalese caves, said to be homes of the yeti, the *Brighu Samita*, a magical book written in ancient Sanskrit, purported to hold answers to life's great mysteries, and personal audiences with the Dalai Lama, the holy keeper of the Buddhist religion. Like so many pilgrims before him, Tom Slick was searching for his own "holy grail" of sorts; India and its neighboring kingdoms seemed a likely place to find it.

> *June 25, 1956*
>
> *Dear Maharaja and Princess Usha,*
>
> *Please forgive me for being so long delayed in writing to thank you for the wonderful hospitality that you showed me on my trip to India in February. The visit with you, complete with tiger hunt and stories of the Yeti, was literally the "high point" of my whole trip.*
>
> *The two of you could not have been any nicer to Cathy and me and we had an absolutely marvelous time with you. I will never forget it . . . it will lead me to the Himalayas next year.*

Cathy McLean and Peter Byrne met in New Delhi, during the summer of 1956. She learned that he had been in the RAF until 1947 and had joined a British tea company with estates in Northern Bengal after the war. He also guided professional tiger hunts along the border of Nepal to offset the boredom of planting tea.

He told Cathy he had recently driven to Sikkim in his ancient Austin Seven and trekked with two Sherpas from Gangtok toward Zemu Glacier, where N. A. Tambazi, a Royal Geographic Fellow, had reported a Yeti sighting more than twenty years ago. After finding a single footprint, in the hard snow of a frozen pool near the glaciers, Byrne's Sherpas were anxious to leave the area, sure that a *shookpa*, another Nepalese word for a Yeti, was nearby. He described their departure, laughing. "We completed the last lap of the trek, some twenty-six miles from Sandakphu to Darjeeling, from 10,000 feet down to 2,000 feet and up again to 8,000 feet in one long day." It was obvious that Peter Byrne was fit enough for the strenuous mountain trek and that he was

Slick's first Yeti hunt began on the Arun River, moving up the valley toward Mount Makalu in northeast Nepal. *Peter Byrne Photo Collection*

as intrigued by the Yeti as Tom Slick was. Cathy sent her "approval" of the new expedition member to Texas.

"Peter Byrne was as cool as James Bond," Chuck Slick remembered more than five decades later. "When we were kids, he was a big game hunter who knew how to survive in a cave in Nepal, not the usual boring grownup friend that most parents have."

In the spring of 1957, Peter Byrne joined Tom Slick in New Delhi and they flew together to Kathmandu, where they gathered supplies and porters. Accompanied by Mr. N. D. Bachkheti, an Indian zoologist, and a liaison officer from Nepal, the small group began their reconnaissance expedition, designed to determine whether Slick should fund a much bigger expedition the following year.

Vehicles were scarce in Nepal in 1957, and the best transportation the group of explorers could find was an ancient bus with brakes that did not always work. They traveled southeast and stopped in a village that was used mostly as a Gurkha recruiting post. As the men began to step out of the brightly colored bus, the brakes released and it began to roll, tossing Tom Slick and several others to the rough pavement. It gained speed and the others jumped off frantically; finally, it crashed into a big wooden house at the bottom of the hill. Fortunately, no one inside was hurt, and the expedition team members suffered only cuts and bruises.

Most of Slick's injuries were to his shins, and the long days of hiking ahead would be particularly painful for him. At the time, he thought only of the mission, never considering that he would not continue. It was only after he returned from Nepal and told the story of the bus accident, that he admitted how difficult the going had been.

When the team reached the lower mountain ranges bordering the Arun Valley, both Slick and Byrne found clues that each would forever cherish. Byrne found a set of large footprints in soft ground in the Chhoyang Khola, at an altitude of about 8,000 feet. He described them as "ten inches in length, five-toed prints of a bipedal creature of considerable weight. They were identical to those I'd found in Sikkim years

Despite injuries to his knees and shins, Slick was an active participant as the 1957 expedition that searched the sacred *beyuls* for the elusive Yeti. *Courtesy of the Mind Science Foundation*

Slick discovered this footprint in the snow at about 10,000 feet and made a plaster cast of it. It remained one of his most prized possessions for the rest of his life. *Courtesy of the Mind Science Foundation*

earlier. I'll never forget the eerie thrill of knowing that a Yeti could be just over the next ridge."

In another location nearby, at about 10,000 feet, Slick found similar prints in snow, approximately thirteen inches long. The men photographed their treasures and made copies using plaster molds. They hurried back to Kathmandu, convinced that a full-scale expedition next year would be worthwhile. Tom Slick couldn't wait to get back to Texas to start planning an all-out effort to capture the Snowman.

From his office in San Antonio he began to pursue a variety of ideas; his daily correspondence is full of letters inquiring about articles he had read and stories he had heard that somehow made his mind leap forward in its quest to find the Yeti.

In a letter to Mr. Russell Cone, in Los Gatos, California, he explored the possibility of using dogs to track it:

Dear Mr. Cone,

I have just read a newspaper article on your hunting dogs making a search for a missing boy in Oregon. I am planning a trip to India and Nepal to search for the "abominable snowman" sometime early next year. I am wondering whether you would have any trained dogs who could hunt these animals. These dogs would be used to try to locate and capture these snowmen in the lower bamboo-forested canyons in the area of Kathmandu (north of), which would be at high altitudes. I would appreciate hearing from you at your earliest convenience.

Thanking you in advance, Tom Slick

Mr. Cone did not respond to what must have seemed like a very odd letter, but Tom Slick was not deterred. Through family friends in Winston-Salem, North Carolina, he located another source for tracking dogs. He wrote to Mr. E. T. Tillman in Elba, Alabama, and arranged to buy "two or three of the 'man dogs' at $200 each"; he asked his secretary to find the best way to ship bloodhounds to Nepal.

Many decades later, stories of the strange dogs brought to the Arun Valley by the Slick Expedition still are told with excitement in remote villages by the porters who are now old men. They imitate the unique howls with eerie precision and laugh with delight at the memory of the small "snow boots" that the dogs wore during the trek.

Slick contacted the director of wildlife management at Cornell University, the American Museum of Natural History in New York, and the Georgia State Game Commission in Atlanta, asking each organization about a newly invented tranquilizer gun being used to capture animals. He explained to all three:

I am planning an expedition into the Himalayas in the very near future to hunt for the Yeti (more popularly known as the Abominable Snowman), and I have been informed that a hypodermic needle of some kind has been used successfully in shooting game and animals.

We, of course, do not want to kill this animal but just capture it and would be interested in having any information you are able to give us on this method.

In early November, Slick contacted Lynn Bollinger at Helio Aircraft Corporation in Norwood, Massachusetts, about using the Helio Courier plane to fly at a low altitude through the forested areas of Nepal. A month later, he decided that a Storch observation plane, which could be purchased in Germany, was an even better idea.

Kenneth MacLeish, editor at *Life* magazine, had met Slick at party in New York and was interested in the "Snowman hunt"; they began to discuss sharing the financial support for the expedition and the exclusive adventure stories and photos that would result. By this point, the expedition budget had climbed to $30,000. It included a reconnaissance plane, tracking dogs, state-of-the-art tranquilizer guns, white hunters, Nepalese government liaisons, scientists, Sherpas, photographers, porters, and supplies. The adventure was a month away.

By mid-January, Slick faced a difficult decision. Despite months of trying to line up a cosponsor to help fund the Yeti expedition, none of his prospects had come through. *Life* magazine had agreed to send its photographer and some fancy new cameras and film, but he realized that the final bill for the four-month trek in the Himalayas would be his alone.

He thought about his father—the first Tom Slick—*before* he became known as the "King of the Wildcatters" with his discoveries of some of the largest oil fields in Oklahoma and Texas. Most people had forgotten his father's first nickname—"Dry Hole Tom"—earned during his earlier oil explorations that were not successful. His namesake chuckled and remembered his father's commitment to risk-taking.

"You'll never succeed unless you try," the older Slick had told his son in 1922, as the two sat together on a riverbank in the woods of western Pennsylvania, weighing the pros and cons of trying to cross what seemed like a raging torrent to a child, using a fallen tree trunk as a bridge.

"Regrets are mostly for the things we wished we'd tried," the oilman told his son. Young Tom had crawled along the tree trunk and crossed the river; he decided to fund the entire Yeti expedition and went home to pack for Nepal.

At the very last moment, just weeks before the expedition team's departure, an unexpected cosponsor decided to put up half the financing for the adventure. During a rather ordinary oil business meeting in Fort Worth, Texas, a wildcatter named Kirk Johnson was so fascinated when Slick described the upcoming expedition that he decided to fund half of it. Later, Johnson would explain that Slick's enthusiasm and certainty about the Yeti were incredibly powerful; he "just couldn't resist."

In early 1958 the Slick-Johnson Snowman Expedition got underway. A highly respected American naturalist named Gerald Russell was field leader. Inspired by his earlier participation on the Daily Mail Expedition in 1955, Russell was convinced the group would find its Yeti in the Himalayas. Professional photographer George Holton and German

The 1958 expedition included (*left–right*) Bryan Byrne, Captain Bahadur Rana, German filmmaker Norman Dyhrenfurth, Peter Byrne, and naturalist Gerald Russell. *Peter Byrne Photo Collection*

moviemaker Norman Dyhrenfurth both would document the journey; Peter Byrne and his brother Bryan would use their superb hunting skills and knowledge of the mountains to track the elusive Yeti. Captain Pushkar Shumsher Jung Bahadur Rana was appointed by the government of Nepal to accompany the expedition. Ten Sherpa guides and sixty-five porters were also engaged for the expedition, and Byrne set about arranging for food and supplies in Kathmandu.

"Food was difficult," he remembered in 2001. "Most of it was canned stuff from India and we ended up throwing a lot away when we opened it. We learned that it was easier to buy fresh food from the villages we passed through, and by the last expedition that's what we did. And of course, our bottle of Scotch was an important addition on all the expeditions."

Breakfast usually consisted of eggs, tea or coffee, and a chapati, the Nepalese version of a tortilla, made from a coarse flour called *ata*. The team members did not stop for lunch, but sometimes snacked on another chapati. Dinner around the campfire might include rice—at the

The Slick-Johnson Snowman Expedition was funded by Slick and Kirk Johnson, a fellow oilman who was captivated by Slick's enthusiasm. *Courtesy of the Mind Science Foundation*

lower altitudes—or potatoes—at the higher altitudes, perhaps roasted chicken or goat, and dal, a hearty lentil soup that is a staple in the mountains. Conversation always turned to the group's mysterious quarry. Sherpas shared the legends they'd heard from their grandparents about the *meh-teh* ("manlike thing that is not a man") or the *dzu-teh* ("big thing"); after a few sips of Scotch, the explorers would go to their tents to dream of a capture just beyond the next mountain or valley.

From base camps that were established in the upper Arun Valley, search parties fanned out to interview Sherpa villagers and search for evidence of the Yeti. Whenever a Sherpa passed along a story about a possible sighting, it took days and sometimes a week to reach that location. And in keeping with Tenzing Norgay's earlier warning, the cunning creature had always moved on. After three months, a Sherpa named Da Temba announced he had seen a Yeti, but Gerald Russell, who was a part of the small search party, did not see it.

From its grand beginnings and large entourage of explorers and would-be explorers, the expedition dwindled in size as the forays from mountain forests to higher peaks became more challenging and temperatures dropped; after four months, only Byrne, his brother Bryan, and a few Sherpas remained in the game. After another five months, the last Yeti hunters trekked slowly back to Kathmandu, using the last of their food supplies, disappointed that the last nine months had not produced more evidence.

The next year, only Peter and Bryan Byrne returned to the mountains, retained by Tom Slick at the salary of just $100 per month, but driven by something far more compelling than financial reward. They took no tents and no food, planning to live off the land and sleep like the native Sherpas do, in caves and wood shelters.

Traveling so lightly meant the Yeti hunters could move quickly, like the elusive creature himself; the Byrne brothers covered huge amounts of Nepalese ground during the next nine months. At Pangboche Temple, the lamas claimed to have a sacred Yeti scalp, which looked convincing to the Byrnes, and may still be protected there. Another famous—or infamous—scalp was brought back from Nepal by Sir Edmund Hillary in 1960; it was proven to be a fake, made from the skin of a serow, a goat antelope. Byrne remained convinced that the fake scalp was made by a

Tibetan taxidermist years before to help ease the inter-temple rivalry that existed between the numerous temples and monasteries of Pangboche, Thyangboche, and Kumjung; Hillary believed he retrieved the original scalp, not a copy, and that it was indeed a fake.

While the Slick team chose to leave the sacred scalp in the Pangboche Temple, they could not resist stealing one finger of the sacred Yeti hand that also was displayed there. Legends of the hand had traveled as far as London, and during an expedition-planning session with biologists in England in 1958, Dr. Osmond Hill had encouraged Tom Slick and his team to bring back the thumb for study. To help, Dr. Hill presented them with a human hand, removed from an English cadaver in his laboratory, suggesting they switch the human thumb with the thumb on the sacred Yeti hand.

"I'll never forget looking into the brown paper bag that Osmond Hill brought with him to lunch at Claridge's," Peter Byrne said with a laugh. "I also had to be rather creative when I traveled with the thumb to Nepal; I carried it in an empty film cannister."

When he got back to Pangboche Temple, Byrne performed the tricky operation with the consent of the head lama, with the dim glow of butter lamps as his only light. He switched the thumbs because only that digit would establish the link between the creature and *Homo sapiens*. But the index finger of the Yeti hand fascinated Byrne more. "It was a longer finger and had a very long dark fingernail on it," he remembered, wondering about its significance. "It may still be in Pangboche," he added with a twinkle in his eye.

After switching the thumbs, Byrne went to Calcutta, where he joined actor Jimmy Stewart and his wife, Gloria, for dinner at the Grand Hotel. His description of his adventures at Pangboche Temple enthralled the Stewarts, and they were curious about the scientific ramifications that would result from study of the "Yeti thumb." They agreed to smuggle the stolen appendage into England and hand it over to Osmond Hill. Tests supposedly were conducted, and Hill's early results described in letters to Slick appeared promising, but vague. No scientific data were published, and the ill-gotten "Yeti thumb" disappeared after Hill's death a few years later. Some cryptozoologists like Dr. Richard Greenwell continued to hope the mysterious digit might be rediscovered and subjected to modern DNA technology, which would provide some definitive answers.

Peter Byrne spent his last winter in Nepal with his brother, in a cave in the upper Chhoyang Khola, one of the deep gorges that runs into the headwaters of the Arun River in the northeastern part of the tiny mountain kingdom, not far from where he had found a footprint three years before.

Exhausted and low on food and supplies, the men had encountered deep winter snows and temperatures as low as 45 degrees below zero. One evening, as they sat around their fire inside the cave, a Sherpa runner arrived with a letter from Tom Slick.

Still fit and handsome in 2001, at seventy-five years old, Byrne smiled at the memory of two young adventurers in a cave in the middle of Nepal, greeting a man who had just run 140 miles in four days to bring them instructions. The three-year hunt for the Snowman in the Himalayas was over. Now Tom Slick had a new quarry in site, closer to home, and he wanted Peter and Bryan Byrne to join him in a new search for Bigfoot in the Pacific Northwest.

The Bryne brothers spent three years on Slick expeditions, searching for the elusive Yeti and Bigfoot. *Peter Byrne Photo Collection*

Elegant, modern, and filled with art and exotic treasures from around the world, Slick's new home was designed by well-known architect O'Neil Ford. *Courtesy of Ford, Powell & Carson Architects*

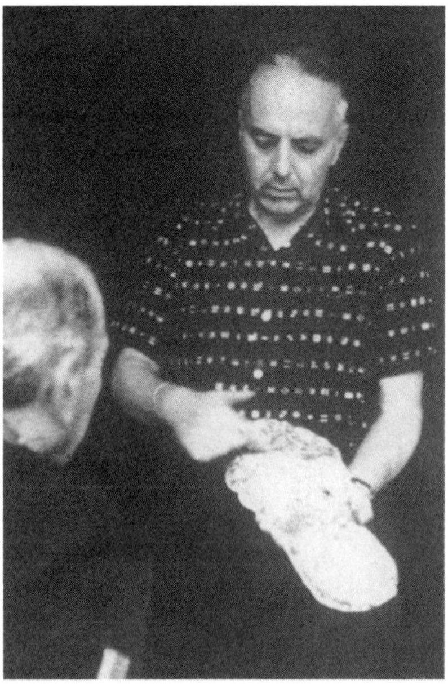

In his new home, Slick's plaster cast of the Yeti footprint was a prized possession. *Courtesy of the Mind Science Foundation*

8

An Open-Door Policy

1956–1961

Between the appetizer course and a feast of roasted pheasant, guests at Tom Slick's dinner party discussed the long table's unusual centerpiece—a large plaster cast of a Yeti footprint. "A mind scientist, a financier, a big game hunter and a beautiful babe were all there," Slick's brother Earl remembered. "It was a typical gathering at Tom's—diverse friends and something interesting to talk about. There was a lively discussion about whether the footprint was real, and bets were made that a Yeti finally would be captured before the decade of the 1960s began."

A waterfall cascaded down one wall of the dining room, and paintings by Pablo Picasso, Georgia O'Keeffe, Georges Braque, and a chimpanzee named Betty dominated the walls. Built by renowned Texas architect O'Neil Ford, whose award-winning designs included Trinity University in San Antonio, Skidmore College in New York, and the famous Santa Maria Beach Club in Lima, Peru, the house was a showcase of contemporary style. A sprawling one-story building with low ceilings and Frank Lloyd Wright–inspired windows, it was surrounded by two acres of lush gardens with a beautiful dark rock swimming pool, in the heart of the city's most affluent neighborhood, just a few blocks from the Urschel estate where his mother and stepfather lived.

Slick and Ford worked well together on the design and building phases of the project, perhaps because the architect genuinely enjoyed his client's constant questions, ideas, and suggestions. In a rather daring move, they decided to use the "Lift-Slab" method of construction

Slick and Ford enjoyed working together; both were full of creative ideas and were not afraid of trying new design and construction techniques. *Courtesy of Ford, Powell & Carson Architects*

developed at the Southwest Research Institute in collaboration with Phillip Youtz in 1951. It was an innovative technique used mainly for multistory buildings, where concrete slabs were poured at ground level, one on top of the other, and then lifted into place on top of columns using hydraulic jacks. There were obvious economic advantages to this method since it used a minimal amount of formwork and allowed year-round construction. It had earned an excellent safety record since much of the work was done on the ground, minimizing the chances of a workman falling on the job.

There were a few instances of failure, however. During construction of Serra High School in San Mateo in 1954, a slab fell due to column instability, and in 1962 a similar accident occurred in Marion, Indiana, when a portion of the roof slab of the girls' dormitory at Marion College collapsed during construction.

Confident that the new construction method was sound and destined for a success in the building industry, Slick and Ford wanted to demonstrate the technique in San Antonio. On May 4, 1958, just a few days before Slick's forty-second birthday, he and Bill Rhame hosted a "Lift-Slab Party" at the new home site at 400 Devine Road.

Friends stood on the slab, which also had an open bar, musicians, and a photographer, as it was lifted to roof level. Guests remembered it as a slow but exciting ride, and Rhame recalled that "the only real drawback was having to climb down a ladder to get to the powder room."

Adjoining the house, garages held Slick's collection of cars that included an elegant green Cadillac and the ultra-sporty vehicle of the times—a Lincoln Continental, with cream-colored interior leather. A motorized miniature Model-T was also parked in the garage, the prized possession of his sons, nine-year-old Tom and seven-year-old Charles.

Always appreciative of beauty and exotic touches, Slick filled the house with art, music, family, and friends. One evening after a dinner party that included parents and children, he invited his guests to view his newest paintings, including one by the legendary artist Pablo Picasso. Some of the adults struggled to describe the abstract images to their children, but they tried—pointing out a girl with a ponytail in the Picasso painting, and a sunset in Slick's newest acquisition. After everyone had returned to the dining room for dessert, they learned the new acquisition had been created by "Betty," one of the chimpanzees housed at the Southwest Foundation. The children were delighted with the joke; the adults were less amused.

Just off the master bedroom, a deep Japanese bathtub opened out to a small private garden with bonsai trees and polished stones. Augusto, the Filipino houseman, kept hot sake near the tub as well as thick terrycloth robes. The ambiance reminded Slick of his adventures in Asia, a world away from San Antonio. Twenty years later, hot tubs would become the rage in the United States, and everyone lucky enough to have seen or sampled Tom Slick's spectacular bathroom would realize they had enjoyed a glimpse into the future.

Although his two marriages ended in divorce, Slick was a very involved father to his four children. When he built his bachelor's home in 1958, its creative nooks and crannies and spirit of fun made it "the place to be." Patty Slick was ten years old when her father's house was completed. She walked into her newly decorated room, discovering lush silks in shades of purple, blue and green, and her own sunken tub in the bathroom. Her father told her that someday she would visit other

Pablo Picasso's painting reminded Slick of his daughter and her ponytail; it became a focal point in the living room of his new home. *Slick Family Photo Collection*

The spirit of fun was ever present in the Slick household, and this painting by a chimpanzee named Betsy was part of his growing art collection. *Slick Family Photo Collection*

places in the world, experience other cultures, and realize that beauty was vastly diverse.

The fabulous fabrics in Patty's room had come from Bangkok, ordered from Jim Thompson, a friend who started the first commercial silk industry in Thailand by employing talented women weavers from the country's more remote northern region of Chang Mai, and allowing them to work at home.

Thompson also was part of the Central Intelligence Agency (CIA), and oblique comments in Slick's correspondence and family stories suggest that both men were collecting information about communist activities and drugs in Southeast Asia at the same time they were each building showplace homes—Thompson's in Bangkok and Slick's in San Antonio.

Slick's years at Yale University coincided with classmates who became OSS agents in the 1940s, graduating to the CIA when that organization was established shortly after World War II. Robin Winks, a Yale history professor who died in 2013, wrote about graduates who joined the OSS in his book *Cloak & Gown: Scholars in the Secret War, 1939–1961*. Harvard professor Ernest May, who specializes in the history of the OSS and CIA, explained that Yale's high involvement in the organizations was rooted in patriotism and bad eyesight.

Shortly before the United States entered World War II, its foreign intelligence organizations were fragmented—the army, navy, Federal Bureau of Investigations (FBI), US State Department, and other entities operated separately, with very little collaboration or communication with one another. In the summer of 1941, President Franklin Roosevelt named William "Wild Bill" Donovan the new Coordinator of Information (COI) and charged him with laying the groundwork for a centralized intelligence agency. The attack on Pearl Harbor a few months later intensified the need for information, and the Office of Strategic Services (OSS) was established with Donovan serving as it first director. He was a perfect choice—a World War I Medal of Honor recipient who had traveled all over the world; he was known for his broad understanding of diverse people and cultures. According to Allen Dulles, a classmate of Slick's at Yale who worked under Donovan, his boss "had a flair for the unusual and for the dangerous, tempered with judgment. In short, he had the qualities to be desired in an intelligence officer."

Donovan is credited with involving Yale in OSS operations as early as 1942, when assistant English professor Joseph Curtiss convinced the university to send him to Turkey, supposedly to collect materials for the Sterling Library, a cover for him to gather sensitive information there.

In a 2014 article for the *Hartford Courant*, William Weir outlined Yale's long history with espionage, noting that "the matter of gathering intelligence isn't far from the skills that make good scholarship. Though 'intelligence' often conjures images of skullduggery and dangerous forays behind enemy lines, there is also the less dramatic matter of research . . . to unearth the important from beneath the trivial."

The Defense Media Network expanded on intelligence as a requirement for espionage, and described the OSS as "Wild Bill Donovan's band of 'glorious amateurs,' comprised of PhD candidates who could win a barfight."

Eventually graduates of elite Ivy League schools and others who joined the OSS or CIA would create a diverse group of patriots that spans multiple generations, including President George H. W. Bush; Allen Dulles, brother of Secretary of State John Foster Dulles; historian and author Arthur Schlesinger; chef extraordinaire Julia Child; Richard Helms; Robert Gates; television journalist Anderson Cooper; and US Representative Will Hurd, to name just a few.

In the 1950s, the hospitality and exotic beauty of Thailand offered a perfect cover for the more sinister activities underway in the Golden Triangle.

But Slick's letter to Thompson in 1956—following a visit to Bangkok —was innocuous for the most part. "One of the very nicest experiences of my trip was to have dinner in your attractive home," he wrote, "and to see your collection of beautiful art and Thai culture. I am sure it gives you great pleasure to watch your house evolve; I am enjoying the same process in San Antonio." He placed an order for silk tablecloths and napkins, specifying that they be "grey or brown or black, with fairly small checkmark designs" for his long dining table and ordered another set for his mother "in a light pink and pastel mixture." At the end of the letter, he urged his friend to be careful.

Slick's concerns proved prescient—a decade later Thompson disappeared mysteriously while visiting friends at Moonlight Cottage in Malaysia's Cameron Highlands and was never seen again. Maxine North, a

Slick's friendship with Thailand's "Silk King", Jim Thompson, resulted in exotic furnishings for his new home, but he worried about the safety of the ex-CIA officer turned entrepreneur. *Courtesy of Texas Biomedical Research Institute*

mutual friend of Thompson's and Slick's, was convinced that the "Thai Silk King" was abducted because of his link to the OSS and CIA, and she continued to search for answers, even offering a cash reward for information, until her death in 2013.

As a starlet at Columbia Pictures in Hollywood, North met and married screenwriter Robert North; they married and moved to Bangkok in 1950, ostensibly to set up the Far East Film Company, a movie distribution firm that was a cover for North's work for the CIA. As the Cold War began, Thailand was considered an important front against communism, and a close—although clandestine—group of intelligence agents was stationed there. Jim Thompson moved to Bangkok the same year and was part of that network.

When her husband died in 1954, Maxine North stayed in Thailand, founding the tremendously successful North Star Company, which manufactured Polaris bottled water, and Thai Celadon, a company that revived the production of a high-quality porcelain style from the same northern region of the country where Thompson's weavers were located. North and Thompson shared a love for Thai art and culture, as well as secrets. They knew Jack Killam, a pilot for Civil Air Transport, who flew weapons and supplies for the CIA from Bangkok to the mountain camps of General Li Mi in Burma, where Chinese troops loyal to Chiang Kai-shek trained; and they knew that loads of opium were flown back to Bangkok in the CIA planes. They were aware of that a project with the code name MKULTRA had been inspired by mind control work in China and Russia, and they supported the evacuation of the Taiwanese Kuomintang (KMT) from Burma, led by the legendary Wild Bill Donovan the year before Robert North died. Previously secret files have been declassified in recent years, and now there is no doubt that Bangkok was a hotbed of spies in the 1950s. Maxine North always believed that Thompson's involvement in this web of intrigue had caused him to vanish.

As the decade of the 1960s approached, Slick continued to move seamlessly between Asia and Texas. There were rumors that he gathered information about the Chinese communists for the CIA, and that he helped organize the Dalai Lama's escape from Tibet to India in 1959, while in Nepal acquiring his much-treasured Yeti footprint.

Everyone wanted to drive eleven-year-old Tom's miniature Model-T car—siblings, friends, and Slick himself. *Betty Slick Moorman Photo Collection*

Letters and diary entries do confirm that he was furnishing a new home with beautiful silks supplied by Thompson. Long before best-selling books began to instruct men to discover their feminine sides, Slick embraced the pleasure he got from selecting lovely accessories for his family and friends, often specifying precise colors and design details. For his daughter's eleventh birthday, he organized "Patty's Pink Pajama Party" and directed his assistant Jeri Walsh to make *everything* pink—from the invitations to the party favors, and of course, all the food that the little girls ate. He surprised the boys with a miniature Model-T car that could travel at a speed of ten miles an hour; and he loved driving it too.

He put the same thoughtfulness into the gifts he ordered and the evenings he planned for numerous girlfriends. Slick's women friends always were attractive and intelligent, and they were scattered around the globe, from Beverly Hills to Beirut, from New York to New Delhi.

Nancy de Herrera was a statuesque blonde widow from California who was spending a year in New York, working as the first "US Ambassadress of Fashion" with legendary newspaper columnist Cobina Wright. Still grieving over the death of her husband—a dashing Argentinian who had played polo, raced cars, and was the country's amateur golf champion—Nancy was hard to please when friends attempted to play matchmaker. But her close friend, Texan Josephine Gill, had known both Nancy and Tom Slick for more than a decade, and she was confident they would like each other. She and her husband had shared many hunting weekends with Slick at various Texas ranches, and they were frequent guests at his dinner parties in San Antonio. "Wear flat shoes," Gill advised her tall friend, "Tom doesn't like women who are taller than he is."

"The first thing I noticed about Tom were those silver-blue eyes," de Herrera recalled many decades later, in her home in Beverly Hills, where she also revealed a long friendship with Maxine North in Bangkok. "Tom and I were relaxed, warm friends right away," she continued. "He talked a lot, like an encyclopedia, but not about personal things. Dating Tom was far from ordinary. He would telephone from Timbuktu to ask me to dinner the following week, and being pursued from some remote corner of the world was quite exciting," she admitted with a smile. Other women friends reported similar calls and letters, and the technique certainly produced far more anticipation and acceptances than the usual single businessman calling from his office downtown.

Something unusual was always a part of time spent with Slick. Before a romantic dinner at an elegant restaurant, he took one date to a pre-dinner appointment with a struggling inventor who was working in his garage to create a seamless tin can. Slick was intrigued; the woman was not as impressed. He gave another a gold-lacquered Buddha hand that she passed down to children and grandchildren in ensuing years.

Family, friends, and even those who met Tom Slick only briefly all were struck by his immense curiosity about everything and his willingness to investigate and explore new ideas. As a result of his ready-to-try-anything personality, his life was high energy and filled with excitement. His children remembered him "always smiling"; business partners described him as an optimist who always looked ahead and did not dwell on disappointments.

Some of his best friends recalled that he was gullible at times. Dr. Robert V. West Jr., who ran the Slick Oil Company, found himself wanting to protect his boss from the constant flow of would-be investment seekers; attorney Jesse Oppenheimer was inspired by Slick's creativity in structuring new deals, but worried that some of the people who pitched those deals were taking advantage of "a man whose door was always open."

Whether he was pursuing the Yeti, or a cutting-edge oil recovery technology, or a new world order that focused on peace, or the best food in town, Slick exuded an energy that was passionate and contagious.

"Tom and I were eating an extraordinary meal at Antoine's Restaurant in New Orleans, with many courses and the elaborate sauces and fine wines that have made that restaurant famous," Bob West remembered, "when the conversation shifted from the oil business to childhood memories. Tom did not often talk about the past, but the topic of the Great Depression came up, and the difficulties that most families experienced in 1929, 1930, and 1931.

"Of course, the Slick family had tremendous, unbelievable wealth, at a time when not many people did. Tom was preparing to go to Exeter Academy and his mother wanted him outfitted properly. She sent the chauffeur to downtown Oklahoma City, with fourteen-year-old Tom in the back seat, instructing him to buy several pairs of new shoes for the year away at school.

"After picking out two or three pairs of shoes, Tom heard a mother and father with a younger boy discover that they could not afford the pair of shoes their son had selected. Quietly, he asked the chauffer, who was preparing to pay for Tom's shoes, to put back a pair and anonymously pay for the other boy's shoes."

Decades later, Bob West's eyes filled with tears as he told the story. "It was such a glimpse of the man," he explained. "His generosity, without any expectation of recognition for it, always touched me deeply."

Apparently, it touched legions. When Slick died in 1962, letters arrived from all over the world; each writer felt that he or she had lost a best friend and each shared a story of some kindness on Slick's part—encouragement, a job, a loan, a special experience. The feeling was mutual; Slick often talked about the importance of friends, calling them "a major source of happiness, along with my family."

In a special audience with His Holiness, the fourteenth Dalai Lama, in Lhasa in 1957, the subjects of both friendship and enlightenment came up. Slick recorded this quote from the spiritual leader in his journal:

> *Whenever I meet people, I always approach them from the standpoint of basic things we have in common. We each have a physical structure, a mind, emotions. Looking at others from this perspective, rather than emphasizing the differences—such as the fact that I am Tibetan, or a different color, or religion, or cultural background—allows me to have a feeling that I am meeting someone just the same as me.*

Beneath the quote, Slick wrote, "I know what he means! When I meet new people, I always approach them from the standpoint of common ground. That way it is easy to find a connection."

According to Slick's journal, despite the political turmoil of the moment, with plots already in place to capture and kill Tibet's revered leader, His Holiness appeared calm and smiling, as if he had all the time in the world to chat about the subjects of unexplained phenomena and consciousness that fascinated the odd American pilgrim who had come so far to see him. He laughingly reminded Slick that westerners had to deal with the constraints of the idea that everything can be explained within the framework of a single lifetime, and the notion that everything must be explained and accounted for.

Slick resonated with the message that not everything could be explained, and he knew with certainty that as soon as "answers" did emerge, new questions would replace the old ones. Slick clasped his hands together in the respectful *namaste* salute and bowed, sharing his dream to find solutions to life's most complex mysteries. His Holiness's eyes twinkled with compassion and humor. "How long can you stay?" he asked.

"One week," Slick replied, and the Dalai Lama laughed gently. Contemplation and meditation with the monks filled the next days in Lhasa, and according to journal entries, the powers of mindfulness and intention were clearly demonstrated, culminating with Slick's description of levitation by the monks. While in the mountain sanctuary, Slick began to envision establishing a new research institute, built alongside the Southwest Foundation and the Southwest Research Institute. It would be an important addition to his original concept of a Science City on the Texas range. He noted in his journal, "Shouldn't science and spirit work together?"

After his spiritual encounter in Lhasa in 1957, Slick returned to the United States and created the Mind Science Foundation. By September 1958, it was chartered as a not-for-profit operating foundation, charged with "scientifically studying the vast potential of the human mind." Dr. Wilford Hahn, a biochemist in the Southwest Foundation's first core group of six scientists, was appointed director of the new organization. Slick also invited Dr. Andrija Puharich, director of the Round Table Foundation in Glen Cove, Maine, to serve as an adviser on extrasensory perception (ESP) research, and Dr. K. T. Jahagirdar, a professor of psychology in Bombay, India, to design a study of mental healing.

Slick asked family members and close friends to serve as the first trustees. His sister, Betty Moorman (who had long ago made her brother promise to investigate the ghostly mysteries of their childhood home) and brother-in-law Lew, his stepbrother Charles Urschel and his wife Betty, his business adviser Bill Rhame, and his attorney Jesse Oppenheimer were the first names on his list. He asked another friend, a Texas/California hybrid, C. V. "Woody" Wood Jr., to play an even bigger role.

Woody was a Texan by birth, but he took his cowboy boots and exaggerated twang to California in 1948, where he helped develop Disneyland in the 1950s, ran McCullough Oil Company in the 1960s and 1970s, and worked on movie company mergers for Warner Brothers in the 1980s. As impressive as his professional credentials were, he needed bigger challenges to satisfy an imagination and curiosity that nearly matched Slick's. But when the two first met at a party in the Beverly Hills Hotel, they seemed worlds apart and a partnership seemed highly unlikely.

Slick was strikingly handsome, intellectual, dressed in a well-tailored suit, quietly studying everyone in the room. Wood was boisterous, gregarious, smoking a cigar, wearing snakeskin boots and a loud sports coat. Despite the very different outward appearances, they soon discovered exciting common ground—a deep belief in the huge, untapped potential of the human mind and a fascination with strange unexplained phenomena. At last Slick had a partner in adventure, with both the curiosity and the financial means to ferret out the mysteries that were so intriguing to both men. And Woody's adopted home state of California seemed to be full of seekers and psychics.

A close friend described Wood's bachelor house as "something out of a Halloween horror film—garish, fun, and truly amazing in its own way. All the shelves in the bar were filled with those little liquor bottles you pick up on airplanes. Woody was like a wonderful circus barker, always enticing his audience to experience the next magic act."

In the late 1950s Woody married Joanne Dru, a beautiful actress who often costarred in western films with John Wayne. She smoothed some of her husband's rough edges and built a sophisticated new home in the posh Truesdale Estates of Beverly Hills, but despite the new elegance, cowboy boots remained on Woody's feet and mystery on his mind.

Slick's children were all preteens in the late 1950s, but they remembered Woody as "one of the greatest characters" they ever knew. Their friendship with him lasted into adulthood. Long after their father's death in 1962, Woody continued to serve as a Lifetime Trustee of the Mind Science Foundation.

It was C. V. Wood who convinced Israeli psychic Uri Geller to come to the United States to be studied at the Foundation in 1972. Geller's spoon-bending demonstrations on the *Tonight Show with Johnny Carson* a few years later led to tremendous public interest and a multitude of research projects that attempted to document "mind over matter" (psychokinesis), extrasensory perception (ESP), and clairvoyance.

Until his death in 1987, Wood continued to contribute funds and advice to the Mind Science Foundation, encouraging its researchers to carry on the mystery hunt that first began at Tom Slick's dinner table, where guests had always had fun with the burgeoning new field of parapsychology.

"We were all at dinner one night," longtime friend Patsy Steves recalled, "when Tom asked everyone for their car keys. He passed them to a stranger—a psychic named Peter Hurkos, who was visiting from California and had never met any of us. The psychic held our keys and told us each something about ourselves; all of his observations were accurate, and it was amazing."

A few years later, the same dinner format was expanded, and the exclusive dining club, the Argyle, became a site for paranormal adventure following a trip Slick made to Washington, DC, in 1957. He arranged meetings with Dr. Harvey Savely, chairman of the Medical Division of the United States Air Force, to discuss the impact that mind science might have on studies underway by the military. Savely told him that because of "vulnerable political situations" the air force could not offer financial support for Slick's new foundation, but he shared some research on a man in Richmond, Virginia, whose entire family seemed to possess "extra retinal vision" (ERV). Slick decided to investigate, and a few weeks later, he brought the Foos family to San Antonio. He planned a special event at the Argyle and invited Hurkos and Puharich to attend as well. The evening's program featured a psychic, an expert in ESP, and the Foos family, which had agreed to demonstrate a remarkable claim—the ability to "see" without the use of their eyes.

An invitation from Tom Slick was highly sought after, and the Argyle party attracted San Antonio's most prominent social set. Beautifully dressed and bejeweled women arrived at the club with their husbands in tow. The men were the city's business elite, and while they were curious and admired Slick's intelligence, civic contributions, and wealth, they were skeptical about his newest research interest.

Slick introduced Puharich, Foos, and Hurkos and the show began. Perhaps most dramatic was the Foos demonstration, described later by Slick in a letter to Dr. Savely:

> *Mr. Foos's young daughter was very carefully blindfolded under our supervision so that we could assume she could see nothing; neither was there the possibility of intelligence between any unblindfolded person and the daughter. After this, the girl demonstrated conclusively that she cannot only 'see' but can see in complete detail, including such things as looking up names and numbers in a telephone book.*

The audience was fascinated, and ESP, unexplained sight, and psychic predictions were topics of conversations on golf courses and at dinner parties for weeks. But Slick was having trouble convincing the scientific community in the United States to take the field of mind science seriously. Never one to be discouraged, he broadened his sights and looked to scientific organizations in Europe for support.

He wrote to the French National Academy of Science, following up on a lead that prominent French banker R. de Lailhacar gave him at a party after a polo match in Jaipur, India. According to the banker, the French Academy was supporting experiments in the field of levitation, something that had long fascinated Slick, who had seen Buddhist monks demonstrate the feat of "mind over gravity" firsthand in Tibet.

When he returned to the United States, he predicted that soon more and more people would master this ability; almost immediately, it appeared his pronouncement had been true. Millions of people had witnessed an ordinary-looking man from California float in midair on national television.

With excitement, Slick immediately wrote to Art Baker, the host of the popular show *You Asked for It*, requesting the performer's name and address. Today, it is doubtful a network would provide that informa-

The search for Sasquatch in the Pacific Northwest offered the perfect family adventure with sons Charles "Chuck" and Tom in 1960. *Slick Family Photo Collection*

tion, but in 1957, Slick obtained it and was soon in touch with Mr. Jim Crabe of Santa Monica, California.

He told Crabe of his interest in levitation and "mind over matter," stressing his desire to study these phenomena scientifically. "We are not asking you to reveal any secrets if there is magic or trickery involved in your act," Slick assured him. "However, if you are performing this feat without mechanical aids and would be willing to give a demonstration in front of a group of scientists, we would be most interested in hearing from you." He thoughtfully promised the levitator that his reply would be completely confidential, with no fear of adverse publicity.

Crabe was an honest man. He replied that he was a magician, which gave Slick an exciting new idea. In all future evening programs to demonstrate the paranormal, he would also invite a magician who would attempt to duplicate the various presentations. This appealed to Slick's curiosity as well as his sense of fairness. The twist was well received, and more than forty-five years later, guests who attended an early Mind Science Foundation program remembered it vividly. The magician as the scientific "control" in parapsychology experiments continued over the years; "the Amazing Randi" became one of the most famous spokespersons for skeptical inquirers, and others have followed, including Martin Gardner, Penn & Teller, and Derren Brown.

As the decade of the 1960s began, Slick's major interest was in mind science, but his fascination with the creature first described by his father as *l'Homme des Neiges* still smoldered. He decided to mount one more expedition to search for the "missing link," this time to the Pacific Northwest, where legends of Bigfoot were gaining momentum. He organized a trip that would be a family adventure, included his sons, nephews, a girlfriend, and longtime partners in exploration, Peter Byrne and Travis Richardson.

"He said, 'I know you are skeptical, Travis,'" Richardson remembered, 'but if I am having the wool pulled over my eyes, I want you there with me.' Many years later, the old rancher was not so sure what they saw. He closed his eyes and described a one-armed man named Red, who owned a very special dog that had been trained to hunt bears. Red claimed the dog was fearless and would run anything, but when they came to tracks on an old logging road, the dog's hackles bristled, and it refused to move down the trail. Everyone froze as the dog howled in fear and ran back to the camp.

"Who knows?" Richardson concluded with a shrug.

During the next few years, others took up the search, most notably John Green and Bob Titmus, whose Pacific Northwest Expedition in 1961 was well documented. Cryptozoologist Richard Greenwell analyzed some of that expedition's data at the University of Arizona and remained convinced of its credibility until his death in 2005.

According to Greenwell, Green reported capturing clear Bigfoot tracks on slide film. He gave the undeveloped rolls to Titmus, who sent them on to Slick in Texas. There was no information about them in Slick's letters or institute files; they remain, like the Yeti thumb, part of a lingering mystery. Author Loren Coleman explored that mystery in 1989, when he published *Tom Slick and the Search for the Yeti*. Peter Byrne, explorer, longtime Yeti hunter, and author of several books including *The Search for Bigfoot: Monster, Myth, or Man*, may have been on the quest the longest. After Yeti expeditions in the Himalayas in the

1950s, he moved to Oregon in the 1960s, established the Bigfoot Research Project; and in 2018, in his nineties, he appeared in *Northwest Report*'s television documentary *Bigfoot Believers*, concluding the interview with, "Every morning I get up I think this is the day I may see Bigfoot."

In April 2019, the Indian army's mountaineering expedition team reported finding a set of large "mysterious footprints" measuring 32 × 15 inches near Makalu Base Camp, above the Arun Valley in Nepal. Photographs posted on the army's official Twitter account were met mostly with skepticism; the army confirmed that the videos and photos had "been handed over to subject matter experts" and that their report was meant to "excite scientific temper and rekindle interest."

In April 2019, the official Twitter account of the Indian Army's Directorate General of Public Information tweeted this picture of what is claimed were Yeti tracks in the snow close to Makalu Base Camp in Nepal. *Twitter/@adgpi*

Slick's enduring interest in world peace included participation in several global conferences and his publication of two books on the subject.
Courtesy of the Mind Science Foundation

9

Visions of the Future

1955–1961

"We prepare for war like precocious giants and for peace like retarded pygmies," Eleanor Roosevelt wrote, quoting her former associate in the United Nations, Canadian Lester Pearson.

Tom Slick liked the quote—he used it in his new book project, and again in the speech he delivered at the World Peace Conference in Paris in 1955. It produced a powerful visual image that he hoped would help drive the point, which was sometimes difficult when discussing the somewhat esoteric concept of peace.

In 1951 Slick self-published his own blueprint for world peace, calling it *The Last Great Hope*. For the next few years he tried to influence his many friends in government and business to join the effort, and he was thrilled when he was invited to address the Paris conference in 1955.

"This may well be the year for a dramatic—perhaps even a decisive—advance in mankind's age-long quest for peace," Slick said to a packed house in France, where the ravages of World War II still were evident. Always an optimist, he was certain that "human beings really want lives of order, unity, and peace." And for the few who had not arrived at this evolved state, he suggested that developing technology would make any future wars even more disastrous, adding a new incentive to move the peace process forward.

"The extreme horror of modern war, the bloodsucking drain of the cost of armaments, and now the specter of atomic and hydrogen destruction of whole populations, with other perhaps worse scientific weapons in the background, seem at last to have brought a worldwide revulsion to these intolerable conditions which set the stage, and condition the minds of men, everyplace, for some brave new program to provide the machinery to prevent war," Slick wrote in 1955.

He was premature. Military spending escalated, and more destructive nuclear weapons were developed. The next decades would experience wars in Vietnam, Central America, Sri Lanka, Kosovo and the Balkan States, the Middle East, and several African nations, and the threat of wars in other regions of the world.

But Slick was not alone in his early peace efforts. He collaborated with Norman Cousins, the brilliant editor of the *Saturday Evening Review*, on his research for another book, *A Permanent Peace*, published by Prentice Hall in 1958, and the two became good friends. Slick respected Cousins's creative mind and ability to use his personal network to effect change. He also was openly admiring of the editor's home life, which appeared as full of projects and interesting people as Slick's, but also had a stability that he yearned for.

"It was really a pleasure to visit you and Ellen at your home last weekend," he wrote, "and I think that you are to be envied your very fine way of life. You have a wonderful place, some wonderful children, and I like the fine, interesting things you do—such as your 'Hiroshima maidens' deal." Cousins had become concerned about the victims of the Hiroshima bombing and set up a fund to help children dealing with its aftermath.

Impressed by the project, in 1954 Slick adopted two Korean foster children through a postwar program called the Foster Parents Plan for War Children. Their names were Park Yong Kyu and Lee Young Sook, and in addition to providing financial support, he regularly corresponded with them until his death in 1962. Unlike many "foster parents" in the program who just sent annual contributions, Slick established a real connection with the boys. He wrote to them about his travels, his dreams, and even the weather in Texas. He sent them magazines about science and his new book about world peace; his letters are an insight into the childlike sweetness that was such a part of the man.

Dear Park,

I was so happy to receive your very nice letter the other day. It certainly sounds like you had a wonderful time during your winter vacation. By your description I imagine Korea looks very pretty indeed with a blanket of snow covering it. You said you drew pictures of the snow scenery. Do you like to draw? I would like to see some of your drawings.

I have been traveling quite a bit lately. This summer my other Children and I are going to spend some time down at the coast swimming, fishing, and just relaxing in the sun.

How is your school work coming along? I know you are trying very hard to do your best and that pleases me very much. Write me again real soon and let me know how you are doing.

Affectionately, Your Foster Dad

As Slick traveled around the world several times in the late 1950s, his mind continuously whirled with new ideas and creative tangents for his existing projects. When he learned that his friend Norman Cousins was preparing to go to Africa to meet Dr. Albert Schweitzer, he wrote to him immediately.

"The greatest impression I've had on my latest round-the-world trip is that uncontrolled population increase is one of the world's URGENT problems . . . a good, cheap, practical technique to control reproduction, if available, would go a long way to solve the problem. Perhaps in Africa they might have a valuable lead.

"If you can find out more on your trip, perhaps from Dr. Schweitzer, and obtain some material, I would arrange for our foundation to test it and the chemical work necessary to turn it into a suitable product."

Schweitzer told Cousins that no such testing was being done in Africa but agreed that birth control would be greatly beneficial to the continent. Just a few years later, the first oral contraceptives hit the market, developed and tested at the Southwest Foundation in Texas. A decade later, the population explosion had gained the general public's attention, and the governments of China, India, Indonesia, and others initiated new education programs, financial incentives, and contraceptive options to tackle this urgent problem.

Slick believed that "most of the world's urgent problems could be solved through Science and Peace," and he understood that financial support was critical to bring about change. "Money is a useful tool," he told Cousins and others, urging them to join him in his unabashed solicitation of funds from anyone he could corner and convince to share his visions for the future.

While Cousins was in Africa, Slick sent him still another suggestion. "An interesting man to see down there would be the 'Diamond King,' Williamson, in Tanganyika. He is reportedly one of the wealthiest men in the world and apparently has no family or plans for his great wealth. I am thinking that if you can meet with him, you might intrigue him with the idea of using his wealth for some important, worthwhile purposes. A bachelor geologist with a huge diamond mine in the middle of Africa might be the source of financing for Peace and Progress through Science!"

The son of the "King of the Wildcatters" did not find anything intimidating about picking up the telephone or his pen to contact the "Diamond King." Partly because he had enjoyed wealth and the access it brings for his entire life, Slick was never embarrassed about asking for what he wanted. Whether he was writing the Indian government about Monsoor fish or David Rockefeller about joining his peace initiative, he was at ease with the "nothing ventured, nothing gained" philosophy that he had learned from his father so many years ago.

This carried over to his personal life as well. Women who dated him during his bachelor days, and those who chose not to for one reason or another, remembered that he never gave up. "Like clockwork, he would check in every few months," one former girlfriend from New York recalled. "I always was hoping I would be an 'exclusive' interest, but Tom was completely honest about his feelings and was a fun, thoughtful friend. Of course, he had friends like me all over the world."

His letters prove it. Extremely courteous, almost formal in their wording, they never alluded to anything improper and were not very personal, almost as if he thought the world might read them someday. Many were responses to letters he had received or notes of thanks for lovely weekends at Texas ranches and far beyond. A Christmas list went to Neiman Marcus, and Slick again asked the store's owner to personally select gifts for many special ladies of the moment. Apparently, Marcus was happy to do so.

Annette, Cathy, Cheryl, Cynthia, Irene, Jeanne, Jeri, Mary, Nancy, Nell, Sandra, Sylvia, Toni, Topsy, and three different women named Helen were on the 1958 holiday list. Packages from Neiman Marcus were shipped overseas to Australia, Beirut, and Paris, and across the United States to Arizona, California, New York, Oklahoma, and all over Texas.

"Every time he called for a date, and I refused because I was seeing someone else, he never was offended or angry," a San Antonio woman recalled many decades later. "But he never stopped calling. The word 'no' was just not in his vocabulary."

His former sister-in-law, Linda Seeligson, agreed. "Success and failure never much concerned him," she explained. "It was the quest that interested him. He was willing to back, with money and prestige, anyone who came to him with an idea, even when others found it 'bogus.'

"I remember a man who claimed he had discovered the essence of life. It was a clear liquid in a glass vial. Extensive time and research were expended on this—but Tom never feared the perceptions of others; he was not afraid of seeming foolish."

This quality of deep self-confidence is rare in our achievement-oriented society, where tremendous value is placed on "fitting in" and social approval. Psychologists agree that many of their clients are drawn to therapy by their fears of rejection, failure, and appearing foolish. And anyone who has ever raised a hand to answer a question in a schoolroom, invited someone to the prom, applied for a job, asked for a raise, made a speech, or walked into a room dressed "wrong" probably has felt that flutter of apprehension, pounding heart, and shiver of fear.

How did Slick escape it? Perhaps it started with his childhood lesson in the Clarion woods, when his father encouraged him to defy fear by crossing the river on a fallen log. Perhaps it was the early death of his father, forcing Slick to become the man of the family at just fourteen years old. Perhaps it was overcoming that first hard year at Exeter, resulting in the transformation from a pudgy teen with pimples into a debate champion and top scholar. Liberated from the fear of failure and the limitations imposed by the perceptions of others, Slick never wallowed in past disappointments. Instead, he turned his attention to problems at hand, and always looked to the future.

His children were an important part of that future. Devotion to his daughter and three sons is highlighted in the hundreds of letters he wrote to them, about them, and to their mothers and teachers. Juggling his projects and exotic travels with his ex-wives' plans and antics in order to spend maximum time with Bill, Patty, Tom, and Chuck was not easy. But as in other endeavors, he was persistent. Whenever his two remarried ex-wives—Betty Young and Polly Pater—weren't responsive to his plans, he simply called or wrote again and suggested an alternate plan.

Slick shared his love of international travel with his children (*left–right*) Patty, Chuck, and Tom during a trip to the Bahamas. *Slick Family Photo Collection*

A subscription to *Science Digest* and a set of golf clubs for Bill, an electric train (ordered from Macy's in New York for $5.95) for Tom, a Wonder Plane for Charles, a special manicure set and cocker spaniel puppy for Patty—like most fathers, Slick was thoughtful in his gift selections for his children, checking with their mothers to see what they needed and wanted. He enjoyed combing through toy catalogs, and in some cases, he conducted his own research. Before making the puppy purchase, he wrote to Jim Searle, manager of the Lake Placid Club in New York, where the family had spent a summer vacation, for advice.

> *My little daughter is very anxious to have a cocker spaniel puppy but I have never cared for them because so many of them seem dumb. However yours seemed bright and smart and most satisfactory. If this is true, I would appreciate your letting me know where I could obtain one like it. As this would be a Christmas present for her, I would appreciate hearing from you just as soon as possible.*
>
> *Thanks very much and kindest regards, Tom Slick*

Searle answered quickly, reassuring Slick that cocker spaniels could indeed be intelligent, and Patty received a puppy that Christmas.

Tom Slick harnessed his imagination to plan family trips to places that would appeal to children. The Slicks fished at the Texas coast, spent time with their cousins in the mountains of Roaring Gap, North Carolina, shook hands with Mickey Mouse at Disneyland, watched surfing contests in Hawaii, went to the top of the Empire State Building in New York City, and later, went camping in the Pacific Northwest, where Slick was searching for Bigfoot.

In addition to carefully selected gifts and travel—and perhaps more important—he gave his children encouragement and advice despite long separations from them. His regular "love letters" to them told them how important they were, and they are the only letters in the collection that do not mention his diverse projects, theories, ideas for new ventures, or world vision. Instead, Slick focused his vision entirely on the young children who missed him and the future adventures they would share.

Dear Bill,

I certainly enjoyed the visit with you in Dallas the other day. I am real proud of the way you are growing up and developing into a fine young man. Your mother wrote to me the other day that you are making fine progress in school and that she has received fine reports from your teachers.

I was very pleased to hear you say the other day that you think you would be interested in becoming a scientist. I believe that this would be a most interesting career and one in which you could accomplish a great deal of constructive work, which would give you a very gratifying life.

For this summer, your plans to work on your grandfather's ranch sound like they would be most interesting. I am sure you will enjoy it and get a lot out of the experience. However, it is easy to get hurt at that sort of work, and you are a little young for it, so I hope that you will be careful and cautious, particularly until you get used to it. I have sent your saddle to you in care of your Uncle Bill.

We are looking forward to having you come down and visit a little later in the summer, so let me know as soon as you are ready to come down to San Antonio . . . we could all go down to the Coast for a while. Then, we have a camping and fishing trip up in the mountains of Wyoming with your Uncle Lew and Jeff, which will get you back in time to start school. So have a wonderful summer and be careful not to get hurt. We will be looking forward to seeing you a little later on.

Love, Dad

Slick treasured his large, extended family and approached its younger members with a tenderness that had been a part of his nature since childhood. His siblings and younger cousins remembered him as a thoughtful, gentle hero. Arthur Seeligson, his first cousin, was four years younger; his memories of times spent with Slick were precious more than seventy years later.

"Tom's father had just died when my family moved into the Slick house in Oklahoma City," Seeligson explained shortly before his own death in 2001 at age eighty. "My mother was the youngest sister of

Tom's mother, and my father was the attorney for the estate. Back in the 1930s, it was not unusual for families to 'circle the wagons' in times like that."

Fourteen-year-old Tom took his younger cousin hunting and fishing, shared boyhood secrets, and treated him as an equal, earning his lifetime admiration and love. Eventually they married sisters, and while Slick's marriage did not last, everyone would remain close for the rest of their lives.

"It's hard to describe Tom's special charisma," Seeligson said with a wistful smile. "When he was in a room, there was a sort of magical energy around him. He was very present in each moment, and all of us got the benefit of that."

Many others—adults and children—also felt that energy. Nieces and nephews, godchildren, and friends of all ages each had a story about the man who was an interesting combination of Santa Claus, the Pied Piper, and Mr. Wizard, weaving together an array of amazing toys, the elusive promise of adventures in elusive places, and of course, the excitement of science.

Slick was an early proponent of equal opportunity and envisioned a future when all children would have the same chances in the world. In 1953 he wrote to Dr. Harold Vagtborg, the director of his scientific institutes, to share his views.

> *Dr. S. D. Kane, the Principal of Booker T. Washington Elementary School, is a fine Negro scholar and an able man. In my opinion, our institutions will be better off if we can add different races to our staff. With that in mind, several years ago I suggested to Dr. Kane that if he found any outstanding Negro scientists, we would like to consider them.*
>
> *He now has a young man, Charles W. Johnson, who will graduate from MIT in a couple of years with a degree in mathematics. He is interested in coming back to San Antonio and the possibility of a position at the Institute . . . we could be a catalyst for his dream.*

Today, the letter is not extraordinary, and use of the word "Negro" instead of African American, which became preferred long after Slick's death, is not correct for these times. But in context, written in the early 1950s, it is important.

Slick's own dream—rich, complex, and intertwined with thousands of related idea strands—envisioned a future where science would make the world a better place. In an open letter addressed to "the Young People of Texas" and circulated to various schools around the state in the mid-1950s, he called scientific research "the great open frontier of the future." The letter described advances in agriculture ("new" ideas including synthetic fertilizers, hybrid crops, and conservation), communications (television had just been invented), new materials (plastic had never been seen before), and space exploration (Russia had just launched "Sputnik") as examples of scientific progress.

After the famous space launch in 1957, the world took notice; Slick worried that the United States needed a drastic revitalization of its economy to compete. He wrote to friends in high positions, including President Dwight Eisenhower, expressing concerns that the Russians, British, and Germans were ahead in the international race for economic advancement, suggesting that America was dangerously complacent.

"Perhaps forthcoming developments will prove that our conviction that we lead the world technologically is not correct," he predicted. "If this is true we should be the first to know the facts, and if it takes a jolt to wake us up, the presence of the unwelcome terrestrial visitor with its living passenger circling unbidden in our skies, may serve the role of a most useful Paul Revere to warn us of danger in time to take counteracting measures," he wrote in his message to the president of the United States.

Slick's correspondence was always answered, and Eisenhower considered the Slick/Urschel/Seeligson clan to be good friends. Once stationed at the US Army's Fort Sam Houston in San Antonio, "Ike" and First Lady "Mamie" were frequent visitors to their stately homes in Olmos Park. While Slick's imaginative scenario of visitors from another planet did not receive much public comment from the president, the government did begin to amass quite a collection of classified materials on the subject in the 1950s, and in 1958, Eisenhower signed the National Aeronautics and Space Act, establishing NASA to propel America's space program forward.

Back in this stratosphere, Tom Slick remained fervent and articulate about the importance of scientific discoveries. "If we can continue to train young men and women who will advance the physical progress of our civilization through the powerful agency of scientific research, and at the same time add to our civilization the moral discipline necessary to properly utilize the advances, we could move our world forward, further in the next years than in all of mankind's history. We have the instruments for progress in scientific research and, if we can solve our other problems such as justice, maintaining world peace, and other essential moral elements, we have before us a promising future indeed. The possibilities are definitely here—it is up to the men and women of the future to turn these potentialities into realities."

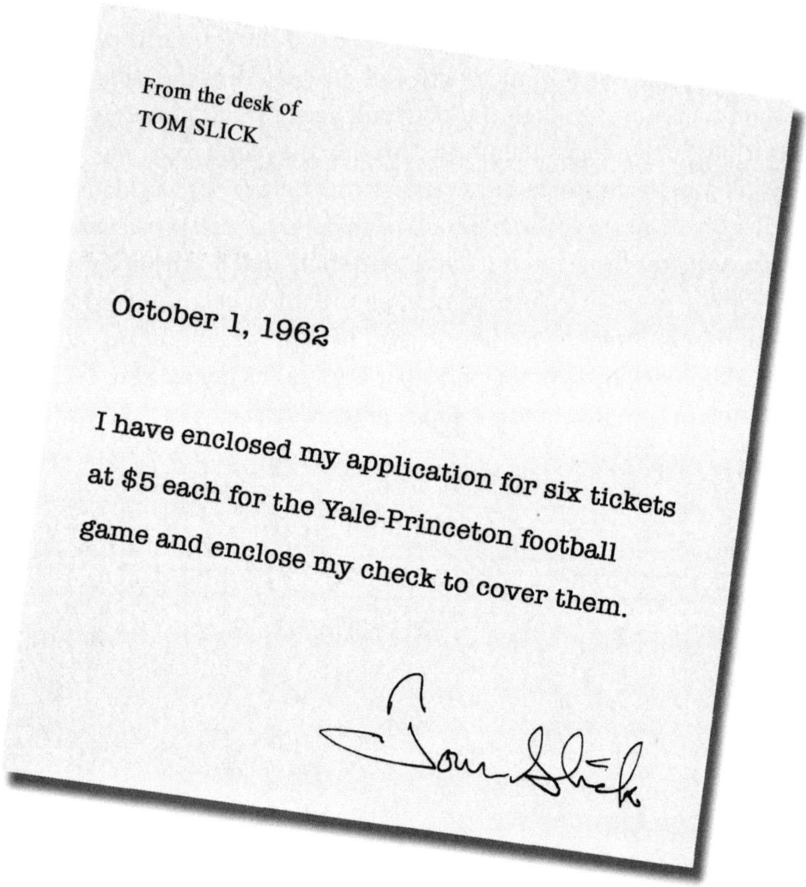

From the desk of
TOM SLICK

October 1, 1962

I have enclosed my application for six tickets at $5 each for the Yale-Princeton football game and enclose my check to cover them.

Tom Slick

As a father who went to Yale, with an oldest son attending Princeton, Slick wondered who the other children would cheer for at the upcoming game between the two Ivy League schools. He purchased six tickets from the Yale Athletic Association on October 1, 1962. *Slick Family Photo Collection*

10

The Unending Search

1962–?

It had been a wonderful week. Tom Slick had joined good friends and oil business colleagues in Calgary, Canada, for some excellent pheasant hunting. The camaraderie and interesting conversations had once again pushed his creative mind into uncharted places, and he was looking forward to pursuing some new projects as a result.

On October 6, 1962, Slick flew south from Canada, toward sunny Arizona, where he planned to have dinner with two of his children—Tom and Chuck—who now lived in Scottsdale with their mother. Then, the following week he would fly to the East Coast, to see the other two—Bill, who was attending Princeton University, and Patty, who was away at Miss Porter's, a boarding school in Connecticut. He had purchased tickets for the Yale/Princeton football game, wondering which team his daughter would cheer, with a "Yalie" for a father and her older brother at Princeton.

As Slick thought about upcoming plans, the weather worsened over Montana. He and the pilot discussed landing and waiting for the storms to end, but they seemed scattered and fear was not a part of Slick's personality. As was his practice, he focused his mind on positive thoughts.

The next four months would be hectic. He planned to squeeze in a quick trip to India and the Himalayas in late October, returning to New York by November 7 to host a major peace conference. In the past week he had written to Nyaya Sharma, Professor Jahagirdar's colleague in Bombay, excited about the new psychic phenomena the man had reported witnessing firsthand. He was eager to learn more; meanwhile, he was practicing the meditation techniques that Sharma had taught him.

"My meditation is going considerably better," he wrote to the Indian, "although I must say that under conditions of my life here I am not carrying it out as regularly as I would like. One interesting thing that has happened is the regular occurrence to me in my mind of the route into your apartment. I absolutely picture the journey, with each turn, from the turn at the corner of the theatre, past the theatre into your street, into your building, into the elevator, out of the elevator on your floor, around to your door and into your apartment, and then into your private room to our sitting place with a clear picture of the group that met there."

While Slick's mind followed the twists of his meditation as the plane lurched, his pilot, Shelly Sudderth, followed the twists of the Beechcraft Bonanza 35 with increased concern. Sometime in the early evening of Saturday, October 6, the plane crashed over Montana's southwestern mountains, about forty miles south of Dillon.

Harold Briggs, the search and rescue coordinator for the Montana Aeronautics Commission, said the plane appeared to have gone to pieces in flight, possibly as a result of an explosion or lightning. Wreckage was strewn over a three-quarter-mile area, and the bodies of Slick and Sudderth were found alongside a country road, nearly a mile from the center of the crash site. Like his father, Tom Slick was forty-six years old when he died.

By Monday morning, news of the accident appeared in the front-page banner headline in his hometown newspaper, the *San Antonio News*; other newspapers and magazines across the country also carried the story. The Slick children had lost a father, people from Texas to Tibet had lost a "best friend," and the world had lost a true mystery hunter. Or had it?

"I plan to work just as hard on the other side," Slick told Jeri Walsh, his longtime personal assistant and sometime girlfriend, and Bill Rhame, his confidant and CEO of Texstar Corporation, both of whom had become ardent supporters of Slick's newest research institute, the Mind Science Foundation. Both Walsh and Rhame were convinced, until their own deaths several decades later, that Tom Slick's quest continued, in another dimension. It definitely continued in this world.

SAN ANTONIO

The Dependable Home Newspaper

NEWS

HOME EDITION ★★★ 5 Cents

45th Year — No. 25 Monday, October 8, 1962 San Antonio, Texas KENS-TV-5 H 32 Pages

Plane Wreckage Examined

TOM SLICK'S DEATH PROBED

Walker Is Back In Texas

Cuba Due To Free Captives

TOM SLICK . . . dies in plane crash

More On The Amazing Career Of Tom Slick On Page 16-B

Montana Crash Kills Pilot, Too

Betty Slick Moorman Photo Collection

With the approach of the twenty-first century and the new millennium, on December 29, 1999, the *Washington Post* announced, "What the millennium holds may be in one's mind," as part of its series *The Next 1000 Years*. According to the cover story, some of the brightest scientists and philosophers in the world today think that the human mind and brain are the final frontier. Slick certainly would have agreed.

As the new millennium began, the Mind Science Foundation (MSF) established the Tom Slick Research Awards for Consciousness to recognize and help fund cutting-edge research in the neurosciences. Recipients have included Belgian scientist Steven Laureys, known for his breakthrough discoveries about the "minimally-conscious" and evidence that happiness can exist there, and David Eagleman, the Stanford University–based researcher best known for his book *The Future of Being Human*. The foundation's educational programming has brought exciting BrainStorm competitions to Texas, featuring scientists from around the world, presenting their newest research breakthroughs in a public competition for MSF funding support. It also has included speakers like

When a paraplegic kicked a soccer ball to open the 2014 World Cup, thanks to exoskeleton technology developed by Dr. Miguel Nicolelis at Duke University, the concept of "mind over matter" was clearly visible to the enthusiastic crowd. *Courtesy of the Mind Science Foundation*

Brazilian researcher Dr. Miguel Nicolelis, who developed the exoskeleton that enabled a paraplegic to kick a soccer ball to open the World Cup in 2014. Tom Slick would have loved this "mind over matter" demonstration, and the fact that science was indeed making the world better.

With new twentieth- and twenty-first-century imaging techniques and collaborations with research labs around the world, Slick's smallest institute moved away from its parapsychology studies into the burgeoning field of neuroscience. In 1999, research on the effects of television violence on young viewers was conducted by John Murray, PhD, MSF's scientist in residence, using magnetic resonance imaging (MRI). The same technology has been used more recently to map the brains of Tibetan lamas during meditation, something Tom Slick practiced in the last few years of his life.

Visionary author and futurist Arthur Clarke (*2001: A Space Odyssey*) suggested that the new millennium would include a human landing on Mars, and that "infrared signals will be detected coming from the center of the Galaxy, the product of a technologically advanced civilization." Slick suggested the same thing in his 1953 letter to President

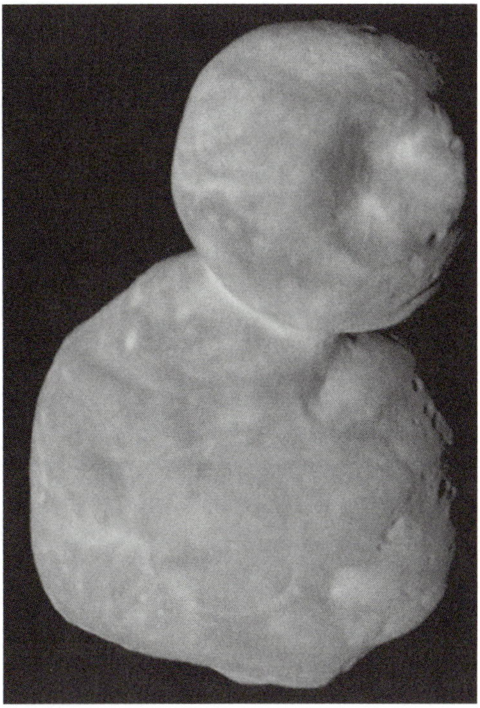

This composite image of Ultima Thule was created from data collected by NASA's *New Horizons* spacecraft and appeared on the May 2019 cover of *Science* magazine. *Courtesy of NASA/JHUAPL/SwRi*

Eisenhower, and followed up with another note in 1957, urging him to establish a space program in the United States. Fast forward to early 2019, when scientists at SwRI created a composite image of Ultima Thule from data obtained by the *New Horizons* spacecraft as it flew past the Kuiper Belt object. Featured on the May 2019 cover of *Science* magazine, the orange-colored space "sculpture" would have thrilled Tom Slick and put to rest his worries that the United States had fallen behind in technological advances.

SwRI has a long history of inventions for air, land, and sea. Its submersibles have been designed to withstand harsh underseas environments, enabling exploration of the RMS *Titanic*, dramatic rescues of sailors caught in disabled submarines, and expeditions that Slick would have joined with gusto. In recent years, marine biologist Sylvia Earle has plumbed the ocean depths for new discoveries, using deep rovers developed at SwRI, and evidence of a giant squid in the deep Mariana Trench off New Zealand provides an adventure every bit as exciting as Slick's search for the Yeti in the Himalayas.

Clarke also predicted that nanotechnology would eventually produce a "universal replicator," allowing us to create any object, however complex, if scientists are given the necessary raw material and information matrix. "Diamonds or gourmet meals can literally be made from dirt," he announced as the new millennium dawned.

Slick liked both diamonds and good food, and new three-dimensional printing technologies have produced both, along with so many other items, ranging from prostheses to guns to complex medical devices.

The medical advances made at the Texas Biomedical Research Institute would no doubt delight Slick, whose first core group of six scientists worked in one primitive laboratory in his ranch house so many decades ago, focusing on agriculture and cattle breeding. In 2019, Texas Biomed employed sixty-five PhD-level scientists and a total of 360 people working in 450,000 square feet of ultramodern facilities, including the state-of-the-art Level Four Biocontainment Laboratory and the Southwest National Primate Research Center, the only place in the world with these combined assets.

Tom Slick first became interested in "apes," a group of primates in which he included chimpanzees, baboons, and orangutans, during a 1952 safari in Kenya with famous white hunter Bunny Allen. He was charmed by the mischievous antics and humanlike expressions of the chimps; at night, several pet monkeys always joined the hunters around the campfire. They reminded Slick of "Johnny," the special surprise he brought home from his trip to South America for his eleven-year-old brother Earl.

As an elegant dinner and drinks were served around the campfire, conversation drifted to the historic lore of ancient Africa, and Allen told Slick that in 1630 a live chimpanzee was supposedly sent by an African chief to the Prince of Orange in the Netherlands. Europe was "horrified and enthralled by the creature," according to reports, and by the end of the seventeenth century, Edward Tyson, one of England's best-known anatomists, wrote that "a chimpanzee resembles Man in many of its Parts, more than any of the Ape-kind, or any other Animal in the world."

Returning from his adventures in Africa, Slick suggested to the scientists at the Southwest Foundation for Research and Education (SFRE) that nonhuman primates might provide a valuable research model for

Baboons were brought to San Antonio from Kenya in 1958, the first members of SFRE's primate research colony; chimpanzees from the Congo arrived a decade later. In 2019, the Southwest National Primate Research Center (*below*) became the largest captive colony of marmosets, dedicated to aging and infectious disease research. *Courtesy of Texas Biomedical Research Institute*

their investigations of heart disease; a few years later, the emerging US space program used a chimpanzee named "Ham" as the first passenger in the Mercury capsule because it was the animal most like humans biologically.

In 1958, SFRE began to build its research colony, with six baboons that were shipped from Kenya. In the early 1960s, Dr. Seymour S. Kalter joined SFRE to study microbiology and infectious diseases and developed the baboon as a model to study viruses; in 1967, the first shipment of chimpanzees from the Congo arrived. Eventually DNA studies established the genetic similarity between humans and primates at 96.4 percent or higher, and by the 1970s and 1980s, primate research had become a vital scientific resource at Texas Biomedical Research Institute, producing breakthroughs in heart disease, neonatal complications, infectious diseases, osteoporosis, cancer, and a wide variety of other human conditions.

The detection of microbes and toxins that could be used in bioterrorism attacks, formation of a tuberculosis (TB) research group, development of a new Hepatitis C therapy, development of experimental therapeutic drugs for the deadly Ebola virus, national collaborations on new ways to detect the mosquito-borne Zika virus, and a $23 million NIH award to work on AIDS vaccine development all have propelled its scientific research forward, attracting new scientists and plans for the future. The Earl Slick Building, designed by the award-winning San Antonio architectural firm Lake Flato, and plans for another Level Four Biocontainment Laboratory are strong indications that Tom Slick's visions for the future are alive and well.

In 2017, physician-scientist Larry Schlesinger, MD, became the president of Texas Biomedical Research Institute and instigated a comprehensive strategic planning process to ensure its continued success going forward. Schlesinger had an impressive background in tuberculosis research, and he organized a research group of fifteen new scientists to study this terrible disease that kills more than 1.5 million people each year. Paving the way for the development of host-directed therapies—therapies designed to rearm and bolster the immune system's ability to fight infection—the approach applies to other infectious diseases under study at Texas Biomed as well. "I believe we can be number one in the country in biocontainment research," he told his staff, board of prominent local and

Larry Schlesinger, MD, became president of Texas Biomedical Research Institute in 2017. *Courtesy of Texas Biomedical Research Institute*

national trustees, and the media, adding that nonhuman primates offer a premier model for preclinical work. "Basic science discoveries can be door-opening discoveries for the clinic, and for the future," Schlesinger explained, an echo of Slick's message so many years ago.

Right next door, the Southwest Research Institute (SwRI) expanded in exciting ways as well. From its modest beginnings at ESSAR Ranch, in the eighteen-room house that Slick called home before moving to nearby San Antonio, SwRI built a staff of two hundred engineers, scientists, and administrators during its first ten years of existence. By 2019, it had 2,608 employees and more than 2.3 million square feet of laboratories and offices on its 1,250+ acre campus. Its revenues had grown to almost $584 million, with commercial projects making up 43 percent of its work and government contracts representing the other 57 percent. Its portfolio of projects would probably stagger even Tom Slick's imagination, "tackling challenges from deep sea to deep space and practically everywhere in between," according to its president, Adam Hamilton.

Slick became interested in a space program in the 1950s, spurred on by Russia's launch of Sputnik. Soon after, SwRI began developing its first rockets (*above*); in March 2000 the first NASA mission led by SwRI was launched from Vandenberg Air Force Base (*right*). *Courtesy of Southwest Research Institute*

As Slick predicted in 1947, the institute has solved problems and developed products that have improved our world in a myriad of ways. Automobile airbags, time-released medicine, polymers for dental appliances, germ-killing paint, Liquid Paper, barcode readers, heart valves, robotics, submersibles, and rocket engines are just a few of the diverse inventions that changed our lives in past decades; since 2018. Since SwRI has been working on new automotive technologies, a rescue submarine for the Australian Navy, advanced "clean" power generation, and earth-orbiting and interplanetary probe missions, to name just a few mind-boggling enterprises underway on what was once just scrub brush northwest of San Antonio.

In 2014, Hamilton became president of Southwest Research Institute. With a background in aerospace engineering, chemistry, biochemistry, and risk assessment, his multidisciplinary expertise has played a major role in the institute's phenomenal growth and diversification in past years. "The Institute continues to be committed to Tom's vision—to solve our clients' toughest challenges day in and day out, year after year, well into the future," he announced during celebrations of SwRI's seventy-first anniversary in 2019.

Adam Hamilton became president of SwRI in 2014. *Courtesy of Southwest Research Institute*

The future was always on Slick's mind, just as it is a favorite topic of the leaders at all the institutes he founded. But the past also holds an important place in the minds of these thinkers; there is recognition that much can be learned from the challenges already faced and conquered (or not yet solved), and from the massive collection of ideas reflected in Slick's correspondence and research notebooks. And of course, there are paradoxes, questions about the greater good, that are such a part of scientific research.

Before the Iraq War began in 2003, the US Army awarded SwRI a six-million-dollar contract to upgrade the CH-47 "Chinook" helicopter fleet. A medium-lift helicopter introduced in 1962, used to haul ammunition, artillery, petroleum, troops, and other cargo for the army, it had developed problems that threatened to ground the fleet. Institute scientists analyzed prior failures and improved the helicopter's reliability in time for use in Iraq in 2003.

The helicopter project would undoubtedly have appealed to Slick's interest in aviation and to his patriotism, but he certainly would have been disappointed to learn that his dreams of worldwide peace were still far from being realized.

Slick's research on the subject occupied much of his time from the middle 1950s until his death. In addition to his two books touting the economic and social advantages of world peace, he created a means to lend his support in the future, by creating the Tom Slick Chair for World Peace at the University of Texas, in Austin, to be funded after his death.

It emerged instead as the Tom Slick Visiting Professor of World Peace, within the Lyndon B. Johnson School of Public Affairs. In 1988, political scientist Brian Urquhart held the position, offering a series of lectures on "Decolonization and World Peace" and later publishing a book with the same title. In 2003, Montgomery C. Meigs, a retired US Army general, was selected by the university to occupy the professorship. Meigs left active duty on January 1, 2003, after a six-year command of all army forces in the ninety-one-country area of responsibility of the US European Command—ironically, just before the Chinook helicopters were redesigned by SwRI in time to enter the Iraq War. Meigs's own experiences during the Vietnam War, Operation Desert Storm, and

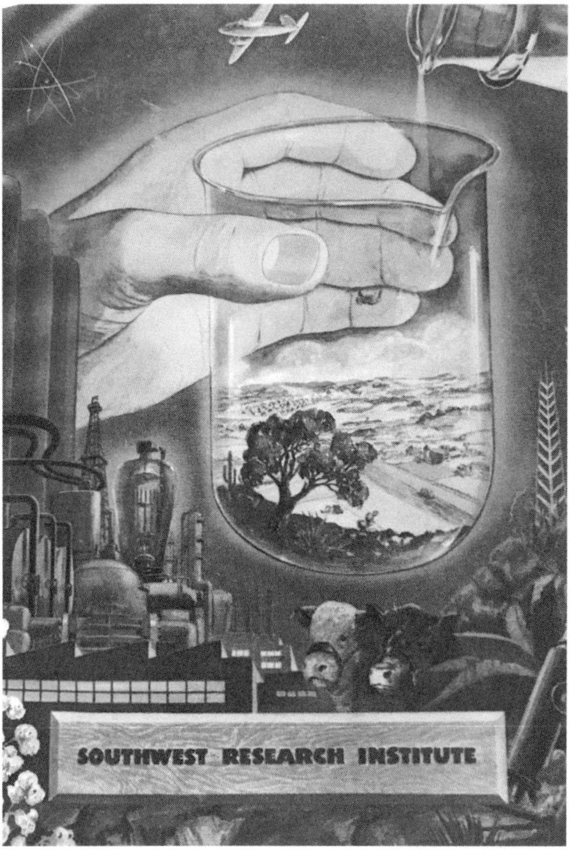

The mysteries of science, and the promise of new discoveries, ignited Tom Slick's curiosity and inspired the creation of a Science City on the Texas prairie. Decades later, this early brochure still evokes that spirit of exploration and the hope for elixirs and solutions that will make the world a better place. *Courtesy of Southwest Research Institute*

two tours of peacekeeping duty in Bosnia made him a powerful voice for articulating the complexities of both war and peace. With the Iraq War underway at the start of the new millennium, it was obvious that sustainable peace efforts still had not become global or a top priority.

Ongoing conflicts in the Middle East, Africa, Chechnya, and several South American countries have demonstrated that world peace, like every mystery that Tom Slick pursued, is elusive. But that never stopped Slick; in fact, he enjoyed the pursuit of anything elusive and increased his efforts when others said things could not be done. His legacy of curiosity and unanswered questions still lives at the institutions he established, and that legacy resonates with new mystery hunters of the twenty-first century and beyond.

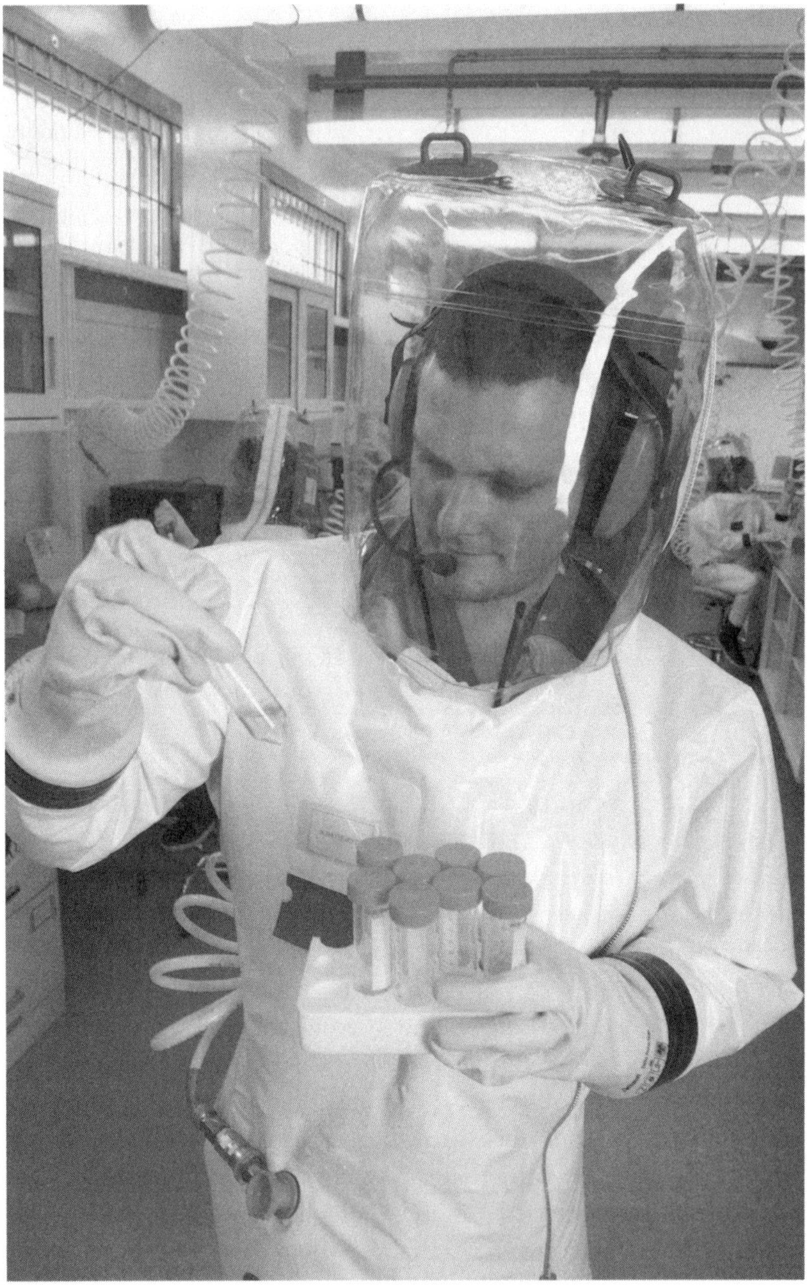

Betty Slick Moorman cut the red ribbon to open the Level Four Biocontainment Laboratory at Southwest Foundation for Biomedical Research in 2000, in honor of her brother Tom Slick. *Courtesy of Texas Biomedical Research Foundation*

Epilogue

The Legacy

In April 2000, applause erupted as a figure in a bright blue spacesuit lumbered over to the thick titanium door of the new Level Four Biocontainment Laboratory and slowly opened it. Inside, other similarly clad scientists would soon study the most terrible infectious diseases of the twenty-first century, hunting down mysterious microbes and pathogens that threaten our lives.

"If Tom Slick were alive, he'd be wearing one of those suits and a grin as big as Texas," a voice in the crowd whispered.

As the microbe hunters modeled their research battle gear, a cool April breeze softened the warmth of the Texas sun. World-famous scientists, board members, journalists, community leaders, and the family of Tom Slick gathered to cut a bright red ribbon, opening the newest laboratory at the Southwest Foundation for Biomedical Research.

The lab would begin its work the following month, in May, and family and old friends of Tom Slick remembered his May birthday and liked this celebration of sorts for a remarkable man who did not live to see his dream reach twenty-first-century proportions.

Betty Slick Moorman walked slowly around the makeshift stage to the waiting ribbon. Still delicately beautiful at eighty-one years old, her eyes glistened slightly as she thought about her older brother, dead for nearly thirty-seven years, whose sometimes far-fetched dreams became realities like today's laboratory. She thought about her own husband, Lewis Moorman Jr., dead only one year, a partner in all her family's endeavors, the steady hand that raised Southwest Foundation to its position as a leading research center in the world. She smiled gently at her younger brother, Earl, seventy-seven years old, from Winston-Salem,

North Carolina, oilman and entrepreneurial founder of Slick Airways, confidant of his brother through every adventure, in Texas today for the dedication. With a special nod to her son-in-law, John Kerr, now chairman of the board of the Southwest Foundation, Betty reached the ribbon. Sunlight illuminated the oversized scissors, and as the blades gleamed, she cut the scarlet banner and welcomed the crowd into the future.

"*I would like to tell you how important the machinery of science is towards the future advancement of our civilization*," Tom Slick wrote in 1952. "*Science gives us a tool of unparalleled effectiveness by which we can improve the physical side of our lives, and since science recognizes no boundaries, the lives of people all over the world*."

Certainly some of the deadliest threats the new lab was preparing to study were known worldwide, most notably in West Africa, South America, Mexico, and Asia. Relative "newcomers" to Western researchers, emerging pathogens like Dengue, Sabia, Ebola, Guanarito, and Hantavirus soon would be captured behind the heavy titanium doors, along with more familiar killers like the HIV virus, which causes AIDS, and Hepatitis C, the terrible enemy of the liver that usually leads to fatal hepatocellular carcinoma. The space-suited team of microbe hunters was ready for them.

Demonstrating their specialized ventilation and waste management systems, full-body suits with respiratory support, emergency backup breathing air tanks connected by spiraling yellow coils, and special showers of powerful disinfectants, they showed off their new protective tools with the rapture of small children on Christmas morning, waking to discover beneath the tree a shiny red miniature fireman's suit, the long-dreamed-of chemistry set, or the newest robotic toy that Santa Claus could find. They smiled at the audience and looked at the lab's shining door with anticipation, energized by the prospect of battling adversaries that would soon be inside.

The Tom Slick "formula" would be at work there shortly, using the same ingredients that launched his Abominable Snowman/Yeti expeditions to investigate reports of evolution's "missing link" in the Himalayas, the Brangus project in 1948 to produce a heartier breed of cattle, the Lift-Slab construction program in 1954 to test a new engineering innovation that literally raised the roof on the building industry, and the "Special Mice" project in the mid-1950s to research early chemotherapies for cancer treatment. Tom Slick's formula always started with an unsolved mystery, a team of experts, a "capture" of sorts, a thorough investigation held to rigorous scientific standards, and finally, a sharing of the results whatever the outcome.

In a 1952 letter "to the Boys and Girls of Texas," Slick described the tremendous advances of science during the first fifty years of the twentieth century, noting, "probably greater advances have been made than the progress previously attained in the whole past ages of history." He predicted still more impressive breakthroughs by the year 2000.

The audience squinted in the Texas sun, thinking about the remarkable discoveries that the last few decades had witnessed in medicine, confirmation of Slick's prediction so many years ago. Many more would occur just inside the doors of this new 34,000-square-foot research building, named the Betty Slick and Lewis J. Moorman Jr. Laboratory in honor of the sister and brother-in-law who had pursued Tom Slick's vision into the new millennium.

Tim Hixon, former chairman of the board, whose uncle had hunted big game with Slick in the 1940s, donated a wing, as did Dr. Burton Grossman, SBC Communications, and United Services Automobile Association (USAA), all generous corporate citizens with a history of supporting scientific research in Texas. A major grant from the National Institutes of Health (NIH), and funding from private donors from San Antonio and beyond—many with personal connections to Tom Slick and Betty and Lew Moorman—started the foundation's march into the future.

As the crowd applauded on that lovely spring day in Texas, anticipating the discoveries that would emerge as the space-suited scientists explored their new mysteries, it glimpsed the dawn of the twenty-first century. From the adjoining primate center, baboons and chimpanzees joined the noisy salute with their own primal shouts, and more universal questions suddenly permeated the day, along with memories of a man who loved mystery and pursued those questions, always daring to journey into the wilderness of his intuition, to that creative place where no one else had ever been. A few in the audience heard the slow hum of a single-engine airplane, and glancing skyward, wondered if Tom Slick might be just beyond the horizon.

Acknowledgments

~ 2020

Creating this revised edition of *In Search of Tom Slick: Adventurer and Visionary* gave me a wonderful opportunity to make some corrections and update information in the first edition, which was published in 2005, and to relive the joy of spending time again with an extraordinary character whom I have always loved and admired.

The longer, original acknowledgments on the next pages recognize the many people who were instrumental to my research and writing between 1999 and 2005, and I remain grateful to all of them. I realize that some have died during the last fifteen years (noted with an asterisk [*])—sending us a powerful message that stories need to be captured before the voices that can share them have left this world. I am so glad that happened for the Tom Slick story.

For this edition, Susan Taylor, who was part of the Nepal trek described in the book's first chapter, helped relocate and scan photos from that great adventure. Special thanks go to Pam and Will Harte for introducing me to Shannon Davies, the remarkable director of Texas A&M University Press, whose enthusiasm made this edition possible; to Thom Lemmons and Patricia Clabaugh at the Press; and to Crystal Hollis at Seale Studios for adding her artistry to some of the page designs with old scrapbook clippings.

Finally, I am grateful to my husband, Geary Atherton, for providing me with a peaceful writing space at our farm in the Texas Hill Country, for sharing his own wonderful boyhood memories of Tom Slick, and for the insights he contributed as my most supportive reader and best friend.

—Catherine Nixon Cooke, 2020

~ 2005

This five-year research and writing project has taken me on an incredible journey, allowing me to spend time with some remarkable people, and to take those roads less traveled, to some of the world's most remote places. I am deeply mindful that this book project has been a privilege and a joy in my life.

Tom Slick's family shared firsthand stories that were essential to the book, and I am especially grateful to Betty Slick Moorman* for many extraordinary hours of interviews and reminiscing, as well as for access to her treasured scrapbooks and collection of family photographs, many of which appear in this book. Thanks also to Earl Slick*, Ramona Frates Seeligson*, Arthur Seeligson Jr.*, Linda Nixon Seeligson, Frates Seeligson*, and Kent Frates for their wonderful insights about a remarkable man, to Margaret Urschel for photos from her Urschel Family collection, and to US Congressman Lamar Seeligson Smith for loaning me his grandfather's copy of the 1934 book written by Joseph Frates (Tom Slick's grandfather) describing Slick's very first international adventure at the age of eighteen. Slick's children (my cousins) Patty Slick Beem, Tom Slick, and Chuck Slick also were generous in their support of my efforts; their childhood memories gave me important glimpses of their father through a different lens.

Tom Slick had friends all over the world, and many contributed interesting and exciting accounts of times they spent with a complex man they admired. Alphabetically, thanks to General Austin Betts*; Peter Byrne; Nancy Cooke de Herrera*; Josephine Gill*; Joseph Goldzieher, MD; Steven Juhazs, PhD*; Bill Mallow, PhD*; James W. Nixon Jr.*, MD; Jesse Oppenheimer, Esq.*; Pushkar Shumsher Jung Bahadu Rana*; Pemmaraju Rao, MD, PhD*; William T. Rhame, PhD*; Travis and Dorothy Richardson*; Burleson Smith, Esq.*; Patsy Steves*; Jeri Walsh*;

Robert V. West Jr., PhD*; and Duncan Wimpress, PhD*, for their stories, and in some cases, for some of the book's historic photographs.

I am grateful to several outstanding publishing professionals—Sandra Martin, Lisa Hagan, and Stephen Topping, who guided me through the publication of the first edition and always respected the book's focus on mystery. Nell Seeligson Smith conducted important background research for the project; and Anson Seale transferred precious photos, some almost one hundred years old, to digital format. Authors (and dear friends) Whitley Strieber, Anne Strieber*, Johan Reinhard, and Jan Jarboe Russell shared their valuable publishing advice and counsel with me.

Institutions founded by Tom Slick added a critical dimension to my research, and the discovery of Slick's letters in a shed at the Southwest Foundation for Biomedical Research enabled me to probe the mind of a visionary thinker. Special thanks to General Frank Ledford*, MD, president of SWFBR until 2005, and to John VandeBerg, the foundation's scientific director, for opening the doors of their offices, laboratories, and storage shed. Scientists at the Southwest Research Institute, some in their eighties and one over ninety years old, supplied firsthand accounts of early discoveries; and Joseph Dial and Beth Costello, executive director and former director, respectively, of the Mind Science Foundation provided glimpses of recent, current, and future research that Slick would love. Prior to Costello's tenure at MSF, I had the privilege to serve as that foundation's executive director, and I am deeply grateful to the organization's board of directors for its strong support of the Tom Slick book project.

Thanks to The Explorers Club, the Mountain Institute, and Himalayan Holidays for nurturing my love of the Himalayas, and to the small team of good friends who accompanied me to Nepal in 2001 to search for the Yeti (and the ghost of Tom Slick): Dan Bennett, Brian Hanson, Rima Lewis, Susan Taylor, and Brad Towne, PhD.

Finally, there are always those very special people, who seem to miraculously intuit the perfect moment for an action that makes all things possible. Betty* and Bob Kelso established the Catherine Nixon Cooke Chair for Exploration at The Mind Science Foundation in 1999; and that generous contribution allowed me to lead the Yeti expedition to the Himalayas in 2001, interviewing the elderly porters who had traveled with Slick nearly fifty years before, which certainly provided a unique perspective to the book. Close friends Kit and Angie Goldsbury also took me to the mountains, to their beautiful home in Aspen, Colorado, when I just couldn't seem to finish the last chapters of the story. Nestled in my fleece in the crisp air, watching bears come up from the river, I felt "word flow" return, just like that joyful last turn on a long trail home.

—Catherine Nixon Cooke, July 2005

Appendix 1

Texas Biomedical Research Institute (Texas Biomed)

Founded 1941

(Formerly Foundation for Applied Research, Southwest Foundation for Research and Education, Southwest Foundation for Biomedical Research)

"Texas Biomed is embarking on a visionary plan to become the world leader in infectious disease research, paving a bold pathway to making the world a less dangerous place."

—Larry S. Schlesinger, MD, President

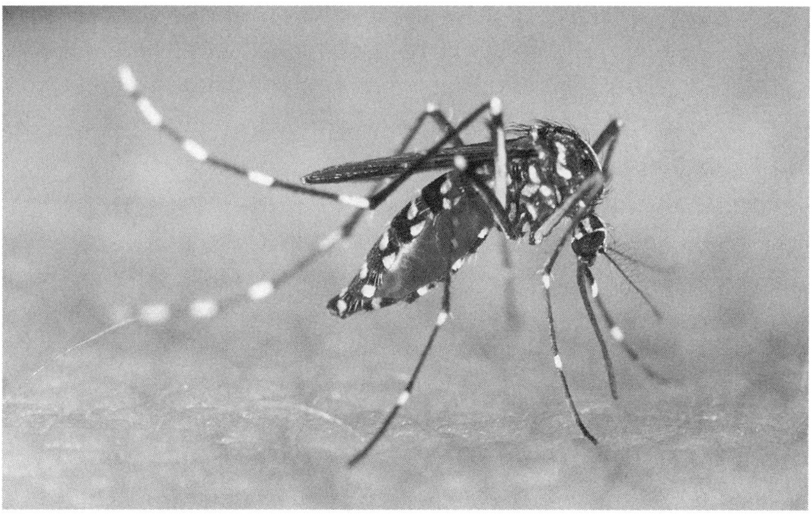

The mosquito-borne Zika virus is being studied at the Texas Biomedical Research Institute. *Courtesy of Texas Biomedical Research Institute*

2017–2019

- SOUTHWEST NATIONAL PRIMATE CENTER became the nation's largest captive colony of marmosets dedicated to aging and infectious disease research, directed by TB/HIV researcher Deepak Kaushal, PhD (2019)
- NEW ZIKA VIRUS AND HIV RESEARCH COLLABORATIONS included an $8 million study to test a promising new AIDS vaccine (2019)
- PROMISING NEW TREATMENT FOR EBOLA TESTED (2018)
- FORMATION OF NEW TUBERCULOSIS (TB) RESEARCH GROUP (2017)

2000–2016

- $23 MILLION FOR AIDS VACCINE DEVELOPMENT AWARDED BY NIH (2016)
- EARL SLICK BUILDING completed with 70,000 square feet of additional laboratories and offices (2014)
- NEW DRUGS AGAINST ANTHRAX AND EBOLA TESTED (2013)
- MALARIA PARASITE GENOME STUDIES INITIATED (2012) with results featured in *Nature* magazine.
- NAME CHANGE/Southwest Foundation for Biomedical Research (SFBR) became Texas Biomedical Research Institute (Texas Biomed) (2011)
- NEW BETTY SLICK AND LEWIS J. MOORMAN OPENED (2000) home to the Burton Grossman Laboratory, the first privately-owned BSL4 in the US

1990s

- 8TH NATIONAL PRIMATE RESEARCH CENTER IN THE US ESTABLISHED AT SFBR (1999)
- GENES INFLUENCING FOOD INTAKE AND BODY FAT ENDOCRINE FUNCTION IDENTIFIED (1997)

- DR. JEAN PATTERSON HIRED AS CHAIR OF VIROLOGY AND IMMUNOLOGY (1996) to build emerging infectious diseases program
- NEW GENETIC SOFTWARE PACKAGE (SOLAR) DEVELOPED BY SFBR (1991)
- GENETICALLY ENGINEERED AIDS VACCINE SUCCESSFUL IN 2 CHIMPANZEES (1990)

1980s

- HIGH FREQUENCY VENTILATOR DEVELOPED AT SFBR in collaboration with SwRI and the US Air Force; tested successfully on ten-week premature chimpanzee (1986)
- NAME CHANGE/Southwest Foundation for Research and Education became Southwest Foundation for Biomedical Research (1984)
- WORLD'S FIRST IN VITRO BIRTH OF NONHUMAN PRIMATE (1983)
- WORLD'S ONLY PRIMATE CENTER WITH HEART STUDY EQUIPMENT (1981)

1970s

- CANCER AND HEPATITIS C RESEARCH AWARDS for chimpanzee research (1978)
- PRESTON G. NORTHRUP LIBRARY dedicated (1971)
- TOTALLY GERM-FREE INFANT BABOONS DEVELOPED AT SFRE (1971)
- SOUTHWEST FOUNDATION FORUM chartered as support group (1970)

1960s

- FIRST SHIPMENT OF CHIMPANZEES FROM THE CONGO ARRIVED (1967)
- HIXON ANIMAL RESEARCH LABORATORIES CONSTRUCTED (1963)

- DR. SEYMOUR S. KALTER JOINS SFRE TO DEVELOP BABOONS AS A MODEL TO STUDY VIRUSES (1963)
- FOUNDER TOM SLICK KILLED IN PLANE CRASH (1962)

1950s

- NIH FUNDING RECEIVED TO STUDY EFFECTS OF HEART DISEASE IN BABOONS (1958)
- FIRST KENYA EXPEDITION TO ACQUIRE PRIMATES (1958)
- BETTY SLICK MOORMAN FOUNDED THE ARGYLE CLUB, where all profits support the work underway at SFRE (1956)
- FIRST BABOON COLONY ESTABLISHED with animals from the local zoo (1956)
- NAME CHANGE/Foundation for Applied Research became Southwest Foundation for Research and Education (SFRE) (1952)

1940s

- FIRST LABORATORY OPENS AT ESSAR RANCH IN THE CABLE HOUSE WITH SIX SCIENTISTS (1947)
- AGRICULTURE AND CATTLE RESEARCH PLANS DEVELOPED (1945)
- FOUNDATION FOR APPLIED RESEARCH (FAR) ESTABLISHED BY TOM SLICK (1941)

Leadership

Larry S. Schlesinger, MD, President (2017–)
Robert W. Gracy, PhD, President (2014–2017)
Kenneth P. Trevett, JD, President (2008–2014)
John C. Kerr, JD, President (2006–2008)
Anthony J. Infante, MD, PhD, President (2005–2006)
Frank. F. Ledford, MD, President (1992–2005)
Duncan Wimpress, PhD, President (1982–1992)
Martin A. Goland, President (1973–1982)
Stanley W. Olson, MD, President (1970–1972)
Harold Vagtborg, DSd, LLD, Administrative Director (1961–1970)
Harold Vagtborg, DSd, LLD, President of FAR (1948–1959)

Appendix 2

Southwest Research Institute (SwRI)

Founded 1947

"Southwest Research Institute has thrived on the Texas ranchland that Tom Slick chose to host this world-class science and engineering organization—it has taken on challenges from the deep sea to deep space and practically everywhere in between."

—Adam L. Hamilton, PE, President and CEO

The *Alvin* submersible was developed by Southwest Research Institute. *Courtesy of SwRI*

Research Highlights

Submersibles

SwRI has a long history designing and building components used in harsh underseas environments. In the early 1960s, it built the pressure hull used in the Alvin Submarine, which was credited with discovering deep-sea vents called "black smokers" and exploring the ruins of the RMS *Titanic.* In 2001, SwRI was selected to build a rescue sub for the US Navy, and in 2019, one for the Australian Navy. In 2013, SwRI redesigned a new, larger hull for the Alvin that allows it to reach 90 percent of the ocean floor.

Robotics

In 2012 SwRI launched the ROS-Industrial (ROSI) Software Repository designed to contain software libraries, tools, and drivers for industrial automation hardware. Over several decades, SwRI has built industrial robots for government and industry, and in 2016 became part of a team building an "agile" robot system that allows the end-effector to be changed depending on the need—allowing the same robot to strip paint from an aircraft, change to an evaluation tool, and then change to a spray painter.

Automotive Technologies

SwRI debuted its patented DEGR (Dedicated-Exhaust Gas Recirculation) engine at the 2014 Society of Automotive Engineers World Congress. A novel engine architecture for passenger cars, DEGR provided an increase in efficiency and performance of up to 15 percent. Being implemented into production vehicles in 2019, it will allow manufacturers to address future fuel economy standards and increasingly stringent emissions levels while providing performance boosts.

Intelligent Traffic Systems

SwRI engineers helped verify and validate San Antonio's advanced traffic management system, TransGuide, which began operations in 1995. Since then, the Institute has developed and applied intelligent transportation systems to more than 13,000 miles of urban and rural roadways in ten states and Puerto Rico.

Autonomous Vehicles

Since the mid-2000s, SwRI has funded an internal research program called the Mobile Autonomous Robotics Technology Initiative (MARTI). This investment in unmanned technologies has mushroomed into numerous advances for government and industry clients, including cooperative vehicles, sensor systems, path planning and military applications.

Advanced Power Generation

The Supercritical Transformational Electric Power (STEP) pilot plant began construction on the SwRi campus in 2019, in partnership with the US Department of Energy/National Energy Laboratory, Gas Technology International (GTI), and GE Global Research (GE). It is a first-of-its-kind 10-megawatt supercritical carbon dioxide (sCO_2) facility that demonstrates the next generation of higher-efficiency, lower-cost electric power technology. Using CO_2 instead of water as the working fluid to generate electricity, overall efficiency as compared to older power plants is increased by ten percent; and the components used are much smaller than those in older plants. Just one desk-sized sCO_2 turbine is capable of powering 10,000 homes.

Earth-Orbiting and Interplanetary Probe Missions

SwRI has built a large array of spaceflight instrumentation going back to the Apollo era when it was contracted to design a fire extinguisher after the Apollo 1 accident in 1967. In the late 1990s, that work took a giant leap forward when SwRI won oversight of its first space mission. Other pioneering missions followed, including the Imager for Magnetopause-to-Aurora Global Exploration (IMAGE) launch in 2000, the New Horizons mission to Pluto in 2006, the Interstellar Boundary Explorer (IBEX) launch in 2008, the Juno mission to Jupiter launched in 2011, and the Magnetospheric Multiscale (MMS) launch in 2015.

Satellite Construction

In 2016 the first satellites built by SwRI launched from Cape Canaveral Air Force Station in Florida. NASA's Cyclone Global Navigation Satellite System (CYGNSS) mission used a constellation of eight identical satellites—all built at SwRI in San Antonio—to peer through thick clouds and heavy rains to measure the movement of ocean waves that are highly correlated to surface-level winds. Still in operation in 2019, the goal of the mission is to provide high-fidelity surface wind estimates as input to tropical cyclone and hurricane models to improve forecasts of hurricane intensification for the first time.

Leadership

Adam L. Hamilton, PE, President (2014–)
J. Dan Bates, President (1997–2014)
Martin Goland, ME, LLD, President (1959–1997)
Harold Vagtborg, DSd, LLD, President (1948–1959)

Appendix 3

Mind Science Foundation (MSF)
Founded 1958

"Tom Slick said it best in 1958: 'I regard the creation of the Mind Science Foundation as the most important undertaking of my life and I plan to devote most of my time to it. I feel that the human mind has tremendous unexplored potential and I want to go about the discovery of that potential in a scientific way.' We at MSF feel the same way today."

—Meriam Good, Executive Director

Mind Science Foundation's BrainStorm Awards (*left–right*: chair of the board, Patricia Smothers, Dr. Subbu Sankarasubramanian, and Meriam Good, executive director). *Courtesy of MSF*

Research and Program Highlights

Neuroscience Research

- In the late 1990s, MSF recognized that new technology could allow scientists to peer inside the brain in exciting new ways. The foundation partnered with Dr. Peter Fox at the University of Texas Health Science Center–San Antonio in a large brainmapping study, using MRI and PET scans for an evolving national database. In 1998, John P. Murray, PhD, became MSF's scientist in residence, using magnetic resonance imagery (MRI) technology to study the effects of television violence on children. His results were widely published and presented at national and international conferences as recently as 2013.
- The Tom Slick Research Awards in Consciousness were created in 2004 to further concentrate on the emerging field of neuroscience, and prominent scientists from around the world have received support. Recipients include Northwestern University's Ken A. Paller, PhD (Improving conscious memory abilities through sleep/2017), Zoran Josipovic, PhD, and Lucia Melloni, PhD, from NYU's Comprehensive Epilepsy Center (Effects of love and compassion meditation on processing in epilepsy/2016), and Alison Gopnik, PhD, and David Whitney, PhD, from the University of California–Berkeley, and Timothy Sweeny, PhD, from the University of Denver (Visual awareness of children/2015). Their research has been published in distinguished scientific journals including *Nature*, *Science*, and the *New England Journal of Medicine* as well as in the popular press (*Time*, *US News and World Report*, and the *New York Times*).
- In 2013, David Eagleman, PhD, became the foundation's chief scientific advisor, charged with guiding MSF's scientific agenda and building a new International Scientific Advisory Board. Eagleman's research has focused on time perception, synesthesia, and sensory substitution—and like Tom Slick, he is an explorer, mystery hunter, and visionary thinker. Based at Stanford University, he is the author of several books including *The Future of Being Human* and the host of the PBS television series *The Brain with David Eagleman*.

Educational Programming

- Since 1984 MSF has sponsored a Distinguished Speakers Series designed to make complex science understandable to a general audience and to offer tools and information for improving human lives. By 2019 audiences had grown from those first small groups of fifty people to standing-room-only crowds of more than four hundred. Topics have ranged from exploring intuition and creativity—subjects Tom Slick pondered—to new research about autism, neurodegenerative diseases, addiction, decision-making, early childhood learning, the teen brain, virtual reality, and more. Speakers have included Jonas Salk, MD, developer of the polio vaccine in 1954; futurist Betty Sue Flowers, PhD, who has guided corporations and individuals in designing their "life plans"; Miguel Nicolelis, MD, a Duke University scientist who developed an exoskeleton that enables movement for the paralyzed; and Michael Platt, PhD, whose research focuses on the neuroscience behind friendship.
- In 2018, MSF joined a Teacher Enrichment Initiative (TEI) partnership with the University of Texas, providing science nights for local schools and STEM instruction for local teachers; a BrainStorm Jr. competition was established in 2019.

MSF-Sponsored Conferences and Competitions

- "International Alzheimer's Disease Symposium," San Antonio, Texas (1985)
- "Mental Fitness for the 21st Century," San Antonio, Texas (1998)
- "International Consciousness Research Retreat," Aspen, Colorado (2008)
- "Understanding Consciousness: Neuroscientific Approaches, Future Directions," Rockland, Maine (2012)
- "Emerging Technologies and the Future of Being Human," San Antonio, Texas (2017)
- "BrainStorm Neuroscience Pitch Competition," San Antonio, Texas (2018, 2019, 2020)

Leadership

Meriam Good, Executive Director (2012–)
Catherine Nixon Cooke, Interim President (2011–2013)
Paul Ingmundson, Executive Director (2009–2011)
Joseph Dial, Executive Director (2003–2018)
Beth Costello, Executive Director (1999–2002)
Catherine Nixon Cooke, Executive Director (1983–1999)
Tom Slick III, Acting CEO (1972–1983)
C. V. Wood, Acting CEO (1962–1972)
Dr. Wilford Hahn, Executive Director (1958–1962)

Bibliography

Primary Sources

Correspondence and diaries of Thomas Baker Slick, 1948–1962. Courtesy of Southwest Foundation for Biomedical Research (now Texas Biomedical Research Institute)

Personal Scrapbooks of Betty Slick Moorman

Interviews conducted by the author, 1997–2001

Betts, General Austin
Byrne, Peter
Cooke de Herrera, Nancy
Frates, Kent
Frates, Rodney
Gill, Josephine
Goldzieher, Joseph, MD
Guenther, Valerie Urschel
Juhasz, Stephen, PhD
Kerr, John C.
Ledford, Frank, MD
Mallow, William, PhD
Moorman, Betty Slick
Nixon, James W., Jr., MD
Oppenheimer, Jesse, Esq.
Rana, Pushkar Shumsher Jung Bhadahu
Rao, Pemmaraju, MD, PhD
Reinhard, Johan, PhD
Rhame, William T., PhD
Richardson, Dorothy
Richardson, Travis

Seeligson, Arthur A. Jr
Slick, Frates
Seeligson, Linda Nixon
Seeligson, Ramona Frates
Slick, Charles Urschel
Slick, Earl
Slick, Jane
Slick, Patty Nixon
Slick, Tom Baker, Jr.
Smith, J. Burleson, Esq.
Steves, Patricia Galt
Urschel, Margaret
Vandenberg, John, PhD
Walsh, Jeri
West, Robert V. PhD
Wimpress, Duncan

Interviews conducted by the author, April 2019

Good, Meriam. Executive Director, Mind Science Foundation
Hamilton, Adam. President, Southwest Research Institute
Miller, Tim. Vice President of Communications, SwRI
Schlesinger, Larry S., MD, President, Texas Biomedical Research Institute

Publications

Apte, Robert Z. *Three Kingdoms on the Roof of the World*. Berkeley, CA: Parallax Press, 1990.
Bailey, W. Scott. "Southwest Research Institute President to Step Down." *San Antonio Business Journal*, April 24, 2014.
Byrne, Peter. *The Search for Bigfoot*. Atlanta: Acropolis Books, 1975.
Coleman, Loren. *Tom Slick and the Search for the Yeti*. Boston: Faber and Faber, 1989.
Davis, Wade. *The Serpent and the Rainbow*. New York: Simon and Schuster, 1985.
Engle, Jeremy. "What Legends and Myths Do You Believe In?" *New York Times*, May 7, 2019.

Euser, Barbara J. *Children of Dolpo*. San Rafael, CA: Writers Center of Marin, 2003.

Garcia, Sandra E. "Yeti Footprints Found in Nepal Are Maybe (Definitely) From a Bear." *New York Times*, April 30, 2019.

Greenwell, J. Richard. "A Sasquatch Field Project in Northern California: Report of the 1997 Six Rivers National Forest Expedition." *Cryptozoology* 13 (1997–98).

Lall, Kesar. *Lore and Legend of the Yeti*. Delhi: Book Faith India, 1988.

Lowe, Sandra. "The Business of Problem-Solving." *San Antonio Business Journal*, February 16, 1997.

Messner, Reinhold. *My Quest for the Yeti*. New York: St. Martin's Press, 2000.

Miles, Ray. *King of the Wildcatters*. College Station: Texas A&M University Press, 1996.

Mind Science Foundation Annual Reports, 2015–19.

Mullin, Glenn H., and Andy Weber. *The Mystical Arts of Tibet*. Atlanta: Longstreet Press, 1966.

Olien, Roger M., and Diana Davids Olien. *Wildcatters: Texas Independent Oilmen*. Austin: Texas Monthly Press, 1984.

Paul, Deanna, and Tania Dutta. "India's Army Claims It Found Yeti Footprints in Nepal: The World Is Skeptical." *Washington Post*, April 30, 2019.

Sakya, Jnan Bahadur. *Short Description of Gods, Goddesses, and Ritual Objects of Buddhism and Hinduism in Nepal*. Kathmandu, Nepal: Handicraft Association of Nepal, 2000.

Sardar-Afkhami, Hamid. "An Account of Padma-Bkod: A Hidden Land in Southeastern Tibet." *Kailash Journal of Himalayan Studies*, 18, nos. 3 and 4 (1996).

Taylor-Ide, Daniel. *Something Hidden behind the Ranges: A Himalayan Quest*. San Francisco: Mercury House, 1995.

Texas Biomedical Research Institute Annual Reports, 2016–19.

Southwest Research Institute Annual Reports, 2016–19.

Twitter ADG-PI-INDIAN ARMY (@adgpi) April 29, 2019.

Vagtborg, Harold. *The Story of Southwest Research Center*. Austin: Southwest Research Institute, distributed by University of Texas Press, 1973.

Index

Note: Page numbers in *italics* indicate photographs and illustrations.